Microbiology: Basic Principles

Microbiology: Basic Principles

Scarlett Jaxon

MURPHY & MOORE
www.murphy-moorepublishing.com

Published by Murphy & Moore Publishing,
1 Rockefeller Plaza,
New York City, NY 10020, USA
www.murphy-moorepublishing.com

Microbiology: Basic Principles
Scarlett Jaxon

International Standard Book Number: 978-1-63987-364-7 (Hardback)

Cataloging-in-Publication Data

Microbiology : basic principles / Scarlett Jaxon.
p. cm.
Includes bibliographical references and index.
ISBN 978-1-63987-364-7
1. Microbiology. 2. Microorganisms. 3. Biology. I. Jaxon, Scarlett.
QR41.2 .M53 2022
579--dc23

TABLE OF CONTENTS

PREFACE

Microbiology is the study of organisms that are unicellular, multicellular or acellular in nature. This comprises bacteria, archaea, viruses, fungi, prions, protozoa and algae, collectively called microbes. These microbes play an indispensable role in processes such as nutrient cycling, biodegradation, climate change, food spoilage, the cause and control of disease, and biotechnology. This field deals with the structure, function, and classification of such organisms as well as the manner in which their activities can be controlled or exploited. In addition to residing in both the inner and outer surfaces of the human body, microbes abound in the soil, in the seas, and in the air. Although normally unnoticed, microorganisms provide ample evidence of their presence. Microorganisms are of incalculable value to Earth's ecology. They are integral to the disintegration of animal and plant remains and converting those into simpler substances that can be recycled in nature. This book unfolds the innovative aspects of microbiology that will be crucial for the holistic understanding of the subject matter. Its objective is to give a general view of the different areas of microbiology, and its applications. This book aims to serve as a resource guide for students and experts alike and contribute to the growth of the discipline.

To facilitate a deeper understanding of the contents of this book a short introduction of every chapter is written below:

Chapter 1- Microbiology is the branch of biology dealing with the scientific study of microscopic organisms. Microorganisms constitute a diverse group of simple life forms that include bacteria, fungi, algae and viruses. It concerns itself with the structure, function and classification of such organisms as well as the research of diseases caused by them, the means to combat such diseases and even their usage in industrial processes. All the diverse aspects of microbiology have been briefly introduced in this chapter.

Chapter 2- Microbial metabolism is the means through which microbes obtain energy and nutrients with the help of chemical reactions, in order to maintain life. Microbial catabolism and anabolism are parts of microbial metabolism, each responsible for the breakdown and buildup of complex molecules. This chapter provides a detailed examination of microbial metabolism, its types and its distinct process.

Chapter 3- Microbial genetics can be defined as the study of genes and gene functions in microorganisms for various purposes. Microbes have relatively simple genetic systems and their study is crucial to the development of molecular and cell biology among others. Bacterial genetics, archaeal genetics, fungal genetics, etc. are some of its aspects which are discussed in detail in this chapter.

Chapter 4- All microbes like other living organisms need food for survival and growth. Their food may contain various forms of nutrients like carbon, oxygen, hydrogen, phosphorus and sulfur. An increase of the microbial population, subject to favorable conditions, is known as microbial growth. Such growth may be induced in controlled scientific environments through culture media. This chapter discusses microbial nutrition and growth for a thorough understanding of the subject.

Chapter 5- There are several theories propounded in the field of microbiology. The Germ Theory of Disease is a widely accepted theory which states that microorganisms can be the cause of certain diseases. The Fermentation Theory and Hologenome Theory of Evolution are two other significant theories. This chapter deals with each theory in a manner that is easily understandable.

Chapter 6- There is no one singular way to undertake the study of microorganisms. Microbiology has many tried and tested methods and techniques which are employed for research. Microbiological culture, bacteriological water analysis, impedance, streaking, thermal shift assay, ATP test, phenotypic testing of mycobacteria, etc. are some microbiological methods and techniques covered in this chapter.

I owe the completion of this book to the never-ending support of my family, who supported me throughout the project.

Scarlett Jaxon

Chapter 1

Microbiology: An Introduction

Microbiology is the branch of biology dealing with the scientific study of microscopic organisms. Microorganisms constitute a diverse group of simple life forms that include bacteria, fungi, algae and viruses. It concerns itself with the structure, function and classification of such organisms as well as the research of diseases caused by them, the means to combat such diseases and even their usage in industrial processes. All the diverse aspects of microbiology have been briefly introduced in this chapter.

Microbiology is the study of microorganisms, which are unicellular or cell-cluster microscopic organisms. This includes eukaryotes such as fungi and protists and prokaryotes such as bacteria and certain algae. Viruses are also included. Microbiology subdivided into divisions including bacteriology, virology, mycology, parasitology and others. A scientist who specializes in the area of microbiology is called a microbiologist. Microbiology can be divided into several subdisciplines, including:

- Microbial physiology: The study of how the microbial cell functions biochemically. Includes the study of microbial growth, microbial metabolism and microbial cell structure.
- Microbial genetics: The study of how genes are organised and regulated in microbes in relation to their cellular functions. Closely related to the field of molecular biology.
- Molecular Microbiology: The study of the molecular biology and genomics of microorganisms.
- Medical microbiology: The study of the role of microbes in human illness. Includes the study of microbial pathogenesis and epidemiology and is related to the study of disease pathology and immunology.
- Veterinary microbiology: The study of the role in microbes in veterinary medicine or animal taxonomy.
- Environmental microbiology: The study of the function and diversity of microbes in their natural environments. Includes the study of microbial ecology, microbially-mediated nutrient cycling, geomicrobiology, microbial diversity and bioremediation. Characterization of key bacterial habitats such as the rhizosphere and phyllosphere.
- Evolutionary microbiology: The study of the evolution of microbes. Includes the study of bacterial systematics and taxonomy.
- Industrial microbiology: The exploitation of microbes for use in industrial processes. Examples include industrial fermentation and wastewater treatment. Closely linked to the biotechnology industry. This field also includes brewing, an important application of microbiology.
- Aeromicrobiology: The study of airborne microorganisms.

- Food Microbiology: The study of microorganisms causing food spoilage.
- Pharmaceutical microbiology: The study of microorganisms causing pharmaceutical contamination and spoilage.

Microorganisms

Microscopic organisms, commonly known as microorganisms or microbes, are found all around us and even inside our bodies. Microorganisms, as their name implies, are so small that they usually require a microscope to see. The vast majority of microbes on the earth pose no real threat to humans, plants or animals; in fact they actually work alongside humans to make world go round, aiding decomposition, decay and even helping us to digest our food. However, there are some microorganisms which negatively impact our lives, causing illness, bad odors and damaging products and surfaces. Microorganisms are found virtually everywhere, except for environments that have been made artificially sterile by humans. Even these must be constantly sterilized and carefully protected, lest microorganisms be tracked in from the outside world. Microorganisms live in water, in soil, and on the skin and in the digestive tracts of animals. This is why all living things must have immune systems – while many microorganisms can be helpful to them, some can be harmful and cause disease. Like all organisms, microorganisms play important roles in the ecosystems they inhabit. Here are a few of their roles. The major groups of microorganisms are bacteria, archaea, fungi (yeasts and molds), algae, protozoa, and viruses. Each type has a characteristic cellular composition, morphology, mean of locomotion, and reproduction.

Bacteria

Bacteria are single-cell organisms that are neither plants nor animals. They usually measure a few micrometers in length and exist together in communities of millions. A gram of soil typically contains about 40 million bacterial cells. A milliliter of fresh water usually holds about one million bacterial cells.

The earth is estimated to hold at least 5 nonillion bacteria, and much of the earth's biomass is thought to be made up of bacteria.

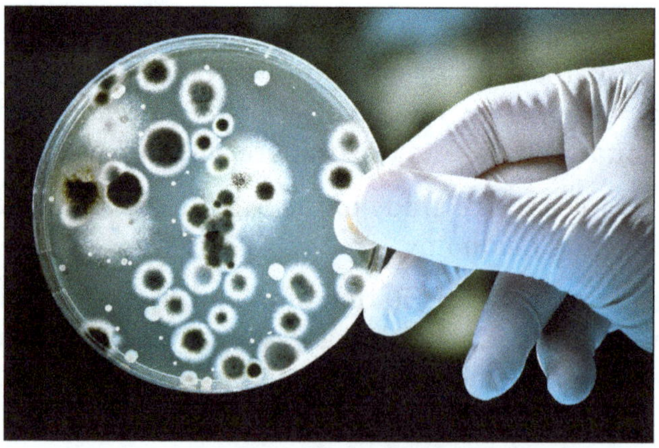

Figure: Bacteria are single-celled organisms.

Types

There are many different types of bacteria. One way of classifying them is by shape. There are three basic shapes:

- Spherical: Bacteria shaped like a ball are called cocci, and a single bacterium is a coccus. Examples include the streptococcus group, responsible for "strep throat."
- Rod-shaped: These are known as bacilli (singular bacillus). Some rod-shaped bacteria are curved. These are known as vibrio. Examples of rod-shaped bacteria include Bacillus anthracis (B. anthracis), or anthrax.
- Spiral: These are known as spirilla (singular spirillus). If their coil is very tight they are known as spirochetes. Leptospirosis, Lyme disease, and syphilis are caused by bacteria of this shape.

There are many variations within each shape group.

Structure

Bacterial cells are different from plant and animal cells. Bacteria are prokaryotes, which mean they have no nucleus. A bacterial cell includes:

- Capsule: A layer found on the outside of the cell wall in some bacteria.
- Cell wall: A layer that is made of a polymer called peptidoglycan. The cell wall gives the bacteria its shape. It is located outside the plasma membrane. The cell wall is thicker in some bacteria, called Gram positive bacteria.
- Plasma membrane: Found within the cell wall, this generates energy and transports chemicals. The membrane is permeable, which means that substances can pass through it.
- Cytoplasm: A gelatinous substance inside the plasma membrane that contains genetic material and ribosomes.
- DNA: This contains all the genetic instructions used in the development and function of the bacterium. It is located inside the cytoplasm.
- Ribosomes: This is where proteins are made, or synthesized. Ribosomes are complex particles made up of RNA-rich granules.
- Flagellum: This is used for movement, to propel some types of bacteria. There are some bacteria that can have more than one.
- Pili: These hair-like appendages on the outside of the cell allow it to stick to surfaces and transfer genetic material to other cells. This can contribute to the spread of illness in humans.

Feeding

Bacteria feed in different ways.

Heterotrophic bacteria, or heterotrophs, get their energy through consuming organic carbon. Most absorb dead organic material, such as decomposing flesh. Some of these parasitic bacteria kill their host, while others help them.

Autotrophic bacteria (or just autotrophs) make their own food, either through either:

- Photosynthesis, using sunlight, water and carbon dioxide.
- Chemosynthesis, using carbon dioxide, water, and chemicals such as ammonia, nitrogen, sulfur, and others.

Bacteria that use photosynthesis are called photoautotrophs. Some types, for example cyanobacteria, produce oxygen. These probably played a vital role in creating the oxygen in the earth's atmosphere. Others, such as heliobacteria, do not produce oxygen.

Those that use chemosynthesis are known as chemoautotrophs. These bacteria are commonly found in ocean vents and in the roots of legumes, such as alfalfa, clover, peas, beans, lentils, and peanuts.

Figure: Bacteria can thrive even in extreme environments, such as glaciers.

Bacteria can be found in soil, water, plants, animals, radioactive waste, deep in the earth's crust, arctic ice and glaciers, and hot springs. There are bacteria in the stratosphere, between 6 and 30 miles up in the atmosphere, and in the ocean depths, down to 32,800 feet or 10,000 meters deep.

Aerobes, or aerobic bacteria, can only grow where there is oxygen. Some types can cause problems for the human environment, such as corrosion, fouling, problems with water clarity, and bad smells.

Anaerobes, or anaerobic bacteria, can only grow where there is no oxygen. In humans, this is mostly in the gastrointestinal tract. They can also cause gas, gangrene, tetanus, botulism, and most dental infections.

Facultative anaerobes, or facultative anaerobic bacteria, can live either with or without oxygen, but they prefer environments where there is oxygen. They are mostly found in soil, water, vegetation and some normal flora of humans and animals. Examples include Salmonella.

Mesophiles, or mesophilic bacteria, are the bacteria responsible for most human infections. They thrive in moderate temperatures, around 37 °C. This is the temperature of the human body.

Examples include Listeria monocytogenes, Pesudomonas maltophilia, Thiobacillus novellus, Staphylococcus aureus, Streptococcus pyrogenes, Streptococcus pneumoniae, Escherichia coli, and Clostridium kluyveri.

The human intestinal flora, or gut micro biome, contains beneficial mesophilic bacteria, such as dietary Lactobacillus acidophilus. Extremophiles, or extremophilic bacteria, can withstand conditions considered too extreme for most life forms. Thermophiles can live in high temperatures, up to 75 to 80 °C, and hyperthermophiles can survive in temperatures up to 113 °C.

Deep in the ocean, bacteria live in total darkness by thermal vents, where both temperature and pressure are high. They make their own food by oxidizing sulfur that comes from deep inside the earth.

Other extremophiles include:

- Halophiles, found only in a salty environment.
- Acidophiles, some of which live in environments as acidic as pH.
- Alkaliphiles, living in alkaline environments up to pH 10.5.
- Psychrophiles, found in cold temperatures, for example, in glaciers.

Extremophiles can survive where no other organism can.

Reproduction and Transformation

Bacteria may reproduce and change using the following methods:

- Binary fission: An asexual form of reproduction, in which a cell continues to grow until a new cell wall grows through the center, forming two cells. These separate, making two cells with the same genetic material.
- Transfer of genetic material: Cells acquire new genetic material through processes known as conjugation, transformation, or transduction. These processes can make bacteria stronger and more able to resist threats, such as antibiotic medication.
- Spores: When some types of bacteria are low on resources, they can form spores. Spores hold the organism's DNA material and contain the enzymes needed for germination. They are very resistant to environmental stresses. The spores can remain inactive for centuries, until the right conditions occur. Then they can reactivate and become bacteria.
- Spores can survive through periods of environmental stress, including ultraviolet (UV) and gamma radiation, desiccation, starvation, chemical exposure, and extremes of temperature.

Some bacteria produce endospores, or internal spores, while others produce exospores, which are released outside. These are known as cysts.

Clostridium is an example of an endospore-forming bacterium. There are about 100 species of Clostridium, including Clostridium botulinum (C. botulinum) or botulism, responsible for a potentially fatal kind of food poisoning, and Clostridium difficile (C. Difficile), which causes colitis and other intestinal problems.

Uses

Bacteria are often thought of as bad, but many are helpful. We would not exist without them. The oxygen we breathe was probably created by the activity of bacteria.

Human Survival

Many of the bacteria in the body play an important role in human survival. Bacteria in the digestive system break down nutrients, such as complex sugars, into forms the body can use.

Non-hazardous bacteria also help prevent diseases by occupying places that the pathogenic, or disease-causing, bacteria want to attach to. Some bacteria protect us from disease by attacking the pathogens.

Nitrogen Fixation

Bacteria take in nitrogen and release it for plant use when they die. Plants need nitrogen in the soil to live, but they cannot do these themselves. To ensure this, many plant seeds have a small container of bacteria that is used when the plant sprouts.

Food Technology

Figure: Cheese making involves bacteria.

Lactic acid bacteria, such as Lactobacillus and Lactococcus together with yeast and molds, or fungi, are used to prepare foods such as cheese, soy sauce, natto (fermented soy beans), vinegar, yogurt, and pickles.

Not only is fermentation useful for preserving foods, but some of these foods may offer health benefits. For example, some fermented foods contain types of bacteria that are similar to those linked with gastrointestinal health. Some fermentation processes lead to new compounds, such as lactic acid, which that appear to have an anti-inflammatory effect.

Bacteria in Industry and Research

Bacteria can break down organic compounds. This is useful for activities such as waste processing and cleaning up oil spills and toxic waste.

The pharmaceutical and chemical industries use bacteria in the production of certain chemicals. Bacteria are used in molecular biology, biochemistry and genetic research, because they can grow quickly and are relatively easy to manipulate. Scientists use bacteria to study how genes and enzymes work. Bacteria are needed to make antibiotics.

Bacillus thuringiensis (Bt) is a bacterium that can be used in agriculture instead of pesticides. It does not have the undesirable environmental consequences associated with pesticide use.

Bacterial species and even specific strains can be differentiated using a number of molecular techniques such as PCR, quantitative PCR, genome sequencing and mass spectrometry. But even without getting into the molecular nitty gritty, there are phenotypic differences between groups of bacteria that can be used to differentiate them. This includes characteristics like their shape (bacilli vs. cocci for example), growth in particular nutrients and preference for high or low oxygen environments. Depending on the characteristic being studied, bacterial species may be broken down into broad groups, but taken together this information can narrow the possible identities greatly. One such useful classification – if a bacterium is Gram positive or Gram negative - is based on the structure of bacterial cell walls.

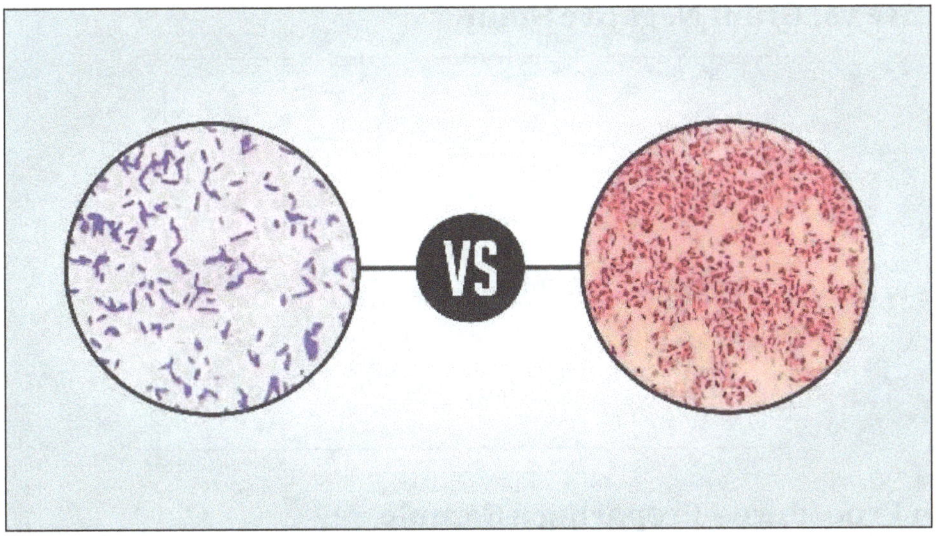

Difference in Structure of Gram Positive vs. Gram Negative Bacteria

The diagram illustrates the differences in the structure of Gram positive and Gram negative bacteria. The two key features that lead to the differing visualization properties of Gram positive and Gram negative species are the thickness of the peptidoglycan layer and presence or absence of the outer lipid membrane. This is because the wall structure affects the cell's ability to retain the crystal violet stain used in the Gram staining procedure which can then be visualized under a light microscope.

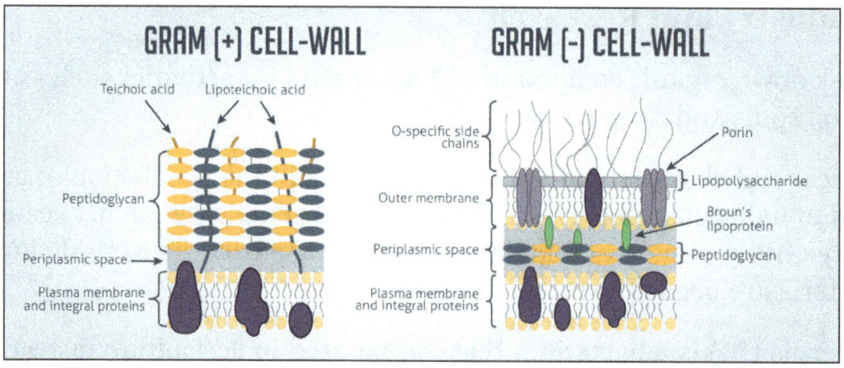

Gram positive bacteria have a thick peptidoglycan layer and no outer lipid membrane whilst Gram negative bacteria have a thin peptidoglycan layer and have an outer lipid membrane.

As Gram positive bacteria lack an outer lipid membrane, when correctly referring to their structure rather than staining properties, are termed monoderms. The outer lipid membrane possessed by Gram negative bacteria means that, when referring to their physical structure, they are termed diderms.

The Gram staining technique was developed in 1884 by Danish bacteriologist Hans Christian Gram. Whilst a Gram stain will not tell you the specific species you are looking at, it can be a quick way to narrow down greatly the list of potential candidates and direct follow-up testing where necessary.

Gram Positive vs. Gram Negative Stain

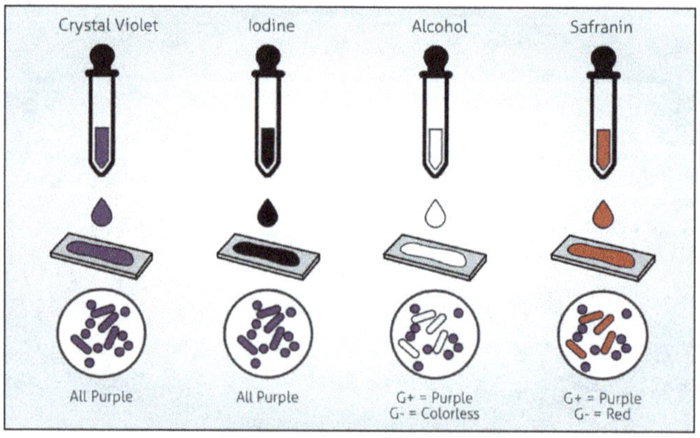

Gram Stain Procedure - Preparing a Sample

- Label a clean glass microscope slide with your sample identification. Ensure you use a pencil as many inks are removed by the reagents used in the staining procedure.
- If preparing your slide from a liquid bacterial culture: Dab a small drop culture onto the slide using a sterile loop. Gently smear the droplet in a circular motion into an area of approximately 1 cm diameter. For very dense cultures it may be necessary to pre-dilute your culture to ensure individual bacterial cells can be seen under a microscope following staining.
- If the source material is from a bacterial plate: Resuspend a loop of colony material in sterile phosphate buffered saline (PBS) and then proceed as for a liquid culture.

- Once the smear has air dried, pass the smeared slide through a flame two or three times. This kills the microbes in the smear and fixes the sample to the slide, be careful not to overheat the sample however as this can distort cellular morphology.

Gram Stain Procedure - Gram Staining a Sample

- Gently flood the smear with crystal violet and leave for 1 minute. Tilt the slide slightly and gently rinse with tap water or distilled water. Crystal violet is a water-soluble dye which enters the peptidoglycan layer in the bacterial cell wall.
- Gently flood the smear with Gram's iodine and leave for 1 minute. Tilt the slide slightly and gently rinse with tap water or distilled water. The smear will now appear purple. Gram's iodine solution (iodine and potassium iodide) is added to form a complex with the crystal violet, which is much larger and is insoluble in water.
- Decolorize the smear using 95 % ethyl alcohol or acetone. Tilt the slide slightly and apply the alcohol drop by drop until the alcohol runs almost clear (5-10 seconds). Immediately rinse with water to avoid over-decolorizing. Decolorizer dehydrates the peptidoglycan layer, shrinking and tightening it. In Gram positive bacteria, the large crystal violet-iodine complexes are then unable to penetrate and escape the thick peptidoglycan layer, resulting in purple stained cells. However, in Gram negative bacteria, the outer membrane is degraded, the thin peptidoglycan layer is unable to retain the crystal violet-iodine complexes and the color is lost.
- Gently flood with safranin counterstain and leave for 45 seconds. Tilt the slide slightly and gently rinse with tap water or distilled water. Safranin is weakly water soluble and will stain bacterial cells a light red, enabling visualization of Gram negative cells without interfering with the observation of the purple of the Gram positive cells.
- Blot the slide dry on filter paper then view the smear using a light-microscope under oil-immersion.

Gram Positive vs. Gram Negative Color

Gram Positive Bacteria

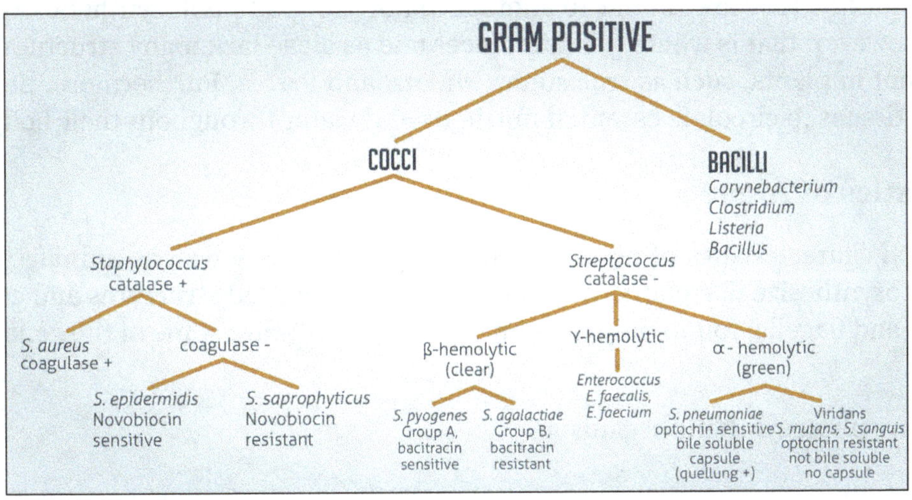

Gram positive bacteria have a distinctive purple appearance when observed under a light microscope following Gram staining. This is due to retention of the purple crystal violet stain in the thick peptidoglycan layer of the cell wall. Examples of Gram positive bacteria include all staphylococci, all streptococci and some listeria species.

Gram Negative Bacteria

Gram negative bacteria appear a pale reddish color when observed under a light microscope following Gram staining. This is because the structure of their cell wall is unable to retain the crystal violet stain so is colored only by the safranin counterstain. Examples of Gram negative bacteria include enterococci, salmonella species and pseudomonas species.

Gram staining cannot be used reliably to assess bacterial phylogenetic relationships. The single membrane of Gram positive species is thought to be the ancestral state. Historically it had been thought that the second membrane found on Gram negative bacteria evolved just once and so all Gram negative species were more closely related to one another than they were to Gram positive species. However, genetic analysis has since shown this not to be the case and it is likely to have evolved multiple times in different lineages - a product of convergent evolution.

Algae

Algae exist in environments ranging from oceans, rivers, and lakes to ponds, brackish waters, and even snow. Algae are usually green, but they can be found in a variety of different colours. For instance, algae living in snow contain carotenoid pigments in addition to chlorophyll, hence giving the surrounding snow a distinctive red hue.

"Alga is a term that describes a large and incredibly diverse group of eukaryotic, photosynthetic life forms. These organisms do not share a common ancestor and hence, are not related to each other (polyphyletic)."

Multicellular examples of algae include the giant kelp and brown algae. Unicellular examples include diatoms, Euglenophyta, and Dinoflagellates.

Most algae require a moist or watery environment; hence, they are ubiquitous near or inside water bodies. Anatomically, they are similar to another major group of photosynthetic organisms – the land plants. However, that is where the differences end as algae lack many structural components typically present in plants, such as true stems, shoots, and leaves. Furthermore, they also do not have vascular tissues to circulate essential nutrients and water throughout their body.

Characteristics of Algae

Specific general characteristics of algae are common to plants as well as animals. For instance, algae can photosynthesize like plants, and they possess specialized structures and cell-organelles, like centrioles and flagella, found only in animals. Listed below are some of the general characteristics of algae:

- Algae are photosynthetic organisms.

- Algae can be either unicellular or multicellular organisms.
- Algae lack a well-defined body, so, structures like roots, stems or leaves are absent.
- Algae are found where there is adequate moisture.
- Reproduction in algae occurs in both asexual and sexual forms. Asexual reproduction occurs by spore formation.
- Algae are free-living, although some can form a symbiotic relationship with other organisms.

Types of Algae

There are many types of algae; however, these are some of the more prominent types:

Red Algae

Also called Rhodophyta, it is a distinctive species found in marine as well as freshwater ecosystems. The pigments phycocyanin and phycoerythrin are responsible for the characteristic red colouration of the algae. Other pigments that provide green coloration (such as chlorophyll a) are present. However, they lack chlorophyll b or beta-carotene.

Green Algae

It is a large, informal grouping of algae having the primary photosynthetic pigments chlorophyll a and b, along with auxiliary pigments such as xanthophylls and beta carotene. Higher organisms use green algae to conduct photosynthesis for them. Other species of green algae have a symbiotic relationship with other organisms.

Members are unicellular, multicellular, colonial, and flagellates. Prominent examples of green algae include Spirogyra, Ulothrix, Volvox, etc.

Not an Alga: Blue-green Algae

In the past, blue-green algae were one of the most well-known types of algae. However, since blue-green algae are prokaryotes, they are not currently included under algae (because all algae are classified as eukaryotic organisms).

Also called cyanobacteria, these organisms live in moist or aquatic environments just like other algae. These include dams, rivers, reservoirs, creeks, lakes and oceans. This class of bacteria obtains energy through the process of photosynthesis. Ecologically, some species of blue-green algae are significant to the environment as it fixes the nitrogen in the soil. Hence, these are also called nitrogen-fixing bacteria. E.g. Nostoc, Anabaena, etc.

However, other types of blue-green algae can be toxic to human beings. They can either be neurotoxic (affects the respiratory or nervous system, causing paralysis) or hepatotoxic (causes the liver to fail). Moreover, some can act as the indicators of environmental health, signaling the extent of pollution.

Examples of Algae

Prominent examples of algae include:

- Ulothrix.
- Fucus.
- Porphyra.
- Spirogyra.

Algal Biofuel

Recent developments in science and technology have enabled algae to be used as a source of fuel. Global demand for petroleum products and declining environmental health has prompted the use of eco-friendly alternatives such as algal biofuel. Hence, algae fuel is an increasingly viable alternative to traditional fossil fuels. It is used to produce everything from "green" diesel to "green" jet fuel. It is similar to the other biofuels made from corn and sugar-cane.

Occurrence of Algae

The algae are ubiquitous (present everywhere) in distribution, i.e., they are found in fresh water as well as marine water, on soil, on rock, as epiphytes or parasites on plants and animals, in hot springs, in desert, on permanent snow-fields etc. But they mainly dwell in aquatic environments.

Based on habitat the algae may be categorized as:

- Aquatic algae.
- Terrestrial algae.
- Algae of remarkable habitats.

Aquatic Algae

Aquatic algae may be fresh water (when salinity is as low-as 10 ppm) or marine (when salinity is 33-40%). Again, certain algae grow in brackish water which is unpalatable for drinking, but less salty than sea water. The fresh water algae usually grow in ponds, lakes, tanks, ditches etc.

The very common fresh water algae are Chlamydomonas, Volvox, Ulothrix, Chara, Oedogonium, Spirogyra, Nostoc, Oscillatoria etc. Some of the very common marine algae are Sargassum, Laminaria, Ectocarpus, Polysiphonia, Caulerpa, Bangia, Padina etc.

Fresh water algae may be termed as planktonic when they grow and remain suspended on the upper part of water (e.g., Volvox, diatom), while the benthic algae are bottom-dwellers. The algae that grow at air-water interface are called neustonic. The benthic algae may be epilithic, that grow on stones; epipelic attached to sand or mud; epiphytic — growing on plants; and epizoic — growing on animal body surface.

The marine algae may be supralittoral or sub- aerial, as they grow above the water level and in the spray zone. The intertidal algae grow in such a depth so that they are exposed periodically due to tides. Other marine algae are sublittoral, meaning that they are constantly submerged at depths as great as 30-60 metres (100-200 ft.).

Again, the supralittoral algae may be edaphic— that grow in and on the soil, epilithic— growing on stones, epiphytic — growing on plants, epizoic— growing on animal body surface, and corticolous — growing on tree barks and parasitic on plants and animals. Some algae (e.g., Chlorella) live endozoically in various protozoa, coelenterates, molasses etc.

Terrestrial Algae

Some algae are found to grow in terrestrial habitats like soils,' rocks, logs etc. The algae that grow on the surface of the soil are known as saprophytes. Many blue-greens, on the other hand, grow under the surface of the soil, and are called cryptophytes.

The algae growing in the desert soil may be typified as endedaphic (living in soil), epidaphic (living on the soil surface), hypolithic (growing on the lower surface of the stones on soil), chasmolithic (living in rock fissures) and endolithic algae (which are rock penetrating). The common terrestrial members are Oscillatoria sancta, Vaucheria geminata, Chlorella lichina, Euglena sp., Fritschiella sp. and Phormidium sp.

Algae of Remarkable Habitats

In addition to above mentioned habitats, some algae also occur in uncommon habitats and termed as:

Halophytic Algae (or Eurhaline)

They grow in the highly concentrated salt lakes, and include Chlamydomonas ehrenbergli, Dunaliella and Stephanoptera sp.

Symbiotic Algae

They grow in association with fungi, bryophytes, gymnosperms or angiosperms. The best examples of symbiotic algae found in association with fungi are Nostoc, Gloeocapsa, Rivularia; the members of Cyanophyceae and Chlorella, Cytococcus, Pleurococcus; the members of Chlorophyceae. This symbiotic association consisting of algae and fungi is called lichen. Nostoc may also associate with Anthoceros and Anabaena associates with the roots of Cycas to form coralloid roots:

- Cryophytic Algae: This group of algae growing on ice or snow provides attractive colours to snow-covered mountains. The alpine and arctic mountains become red due to the growth of the Haemotococcous nivalis; green snow in Europe is due to the growth of Chlamydomonas yellowstonensis. Scotiella nivalis and Raphidonema brevirostri cause black colouration of snow, whereas Ancyclonema nordenskioldii is responsible for brownish purple colouration.

- Thermophytes or Thermal Algae: This group of algae occurs in hot water springs (50- 70°C) where normal life is not possible. Many blue-greens (e.g., Oscillatoria brevis, Synechococcus elongates, Heterohormogonium sp.) are grown in such hot springs.
- Lithophytes: They grow on the moist surface of stones and rocks, e.g., Nostoc,. Gloeocapsa, Enteromofpha, Batrachospermum etc.
- Epiphytic Algae: They grow on other plants including other algal members.

Algae on Algae

- Ptilota plumosa and Rhodymenia pseudopalmatta on Laminaria hyperborean.
- Diatoms on Oedogonium, Spirogyra etc.
- Algae on Bryophytes: Blue-green algae like Nostoc, Oscillatoria, diatoms like Achnanthes etc. grow on different bryophytes.
- Algae on Angiosperms: Algae like Cocconis, Achnanthes etc. grow epiphytically on Lemna, an aquatic angiosperm. Alga like Trentepohlia grows on the barks of different angiospermic plants, and is very common in Darjeeling (India).
- Epizoic Algae: The algae growing on animals like fish, snail etc. are called as epizoic, e.g., Stigeoclonium are found in the gills of fishes.
- Endozoic Algae: They grow in the tissues of animals, e.g., Zoochlorella sp. is found in Hydra viridis.
- Parasitic Algae: Some algae grow parasitically on different plants and animals.

Some parasitic algae are:

- Cephaleuros (Chlorophyceae) is parasitic and grows on the leaves of various angiosperms, such as tea (Camellia sinensis), coffee (Coffea arabica), Rhododendron, Magnolia and pepper (Piper nigrum). The most important one is Cephaleuros virescens, which causes Red rust of tea.
- Rhodochytrium (Chlorophyceae) grows on ragweed (Ambrosia) leaves.
- Phyllosiphon (Chlorophyceae) grows on the leaves of Arisarum vulgare.
- Ceratocolax (Rhodophyceae) grows in Phyllophora thallus.

Psammon

The algae which grow in sandy beaches are called psammon, e.g., Vaucheria, Phormidium etc.

Thallus Organisation in Algae

Thalli of algae show a range of organisation starting from unicellular form to highly organised multicellular habit where the plant body is differentiated into root-like, stem-like, and leaf like structures giving a higher plant-like appearance. Their size ranges from a few microns to several metres.

The algal thalli are grouped into the following, based on their organisation:

- Unicellular Algae: Unicellular forms of algae are also called acellular algae as they function as complete living organisms. Unicellular forms are common in all the groups of algae except Rhodophyceae, Phaeophyceae and Charophyceae. The unicells may be motile or non-motile.
- The motile unicells are either rhizopodial or flagellated.

The rhizopodial forms lack rigid cell wall and have cytoplasmic projections that help them in amoeboid movement, e.g., Chrysamoeba (Chrysophyceae, Rhizochloris (Xantho- phyceae).

The flagellated unicells resemble the motile gametes and zoospores. The flagella function as the organ of locomotion varying in number and type in different groups. The flagellated unicells are found in many groups of algae, e.g., Phacotus and Chlamydomonas, of Chlorophyceae. Euglena of Eugleno phyceae etc.

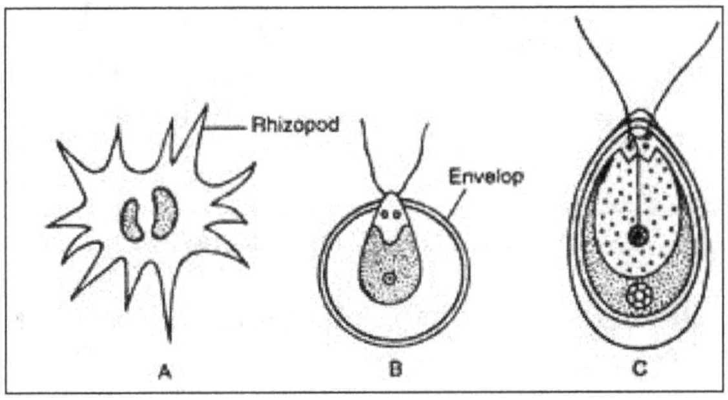

Figure: Unicellular motile algae: A. Chrysamoeba, B. Phacotus C. Chlamdomonas.

- The non-motile cells may be spiral filament as found in Spirulina (Cyanophyceae). The coccoid unicellular algae are the simplest forms of algae found in Cyanophyceae, Chlorophyceae etc., e.g., Gloeocapsa, Chlorella.

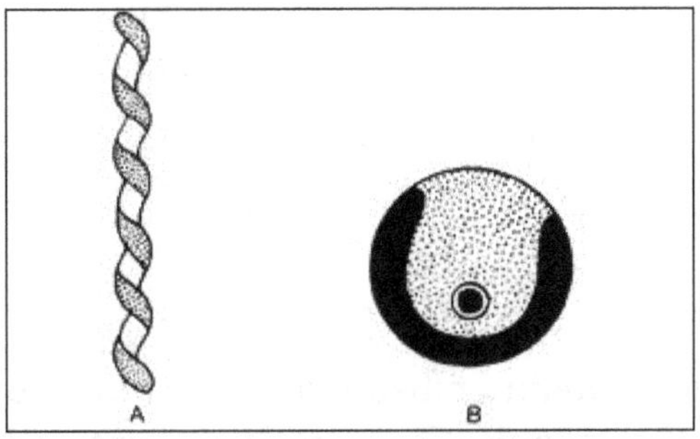

Figure: Unicellular non-motile, A. Spirunila, and B. Chlorella.

Multicellular Algae

- Colonial: The colonial habit is achieved by loose aggregation of cells within a common mucilaginous investment. The cells of these usually remain connected with each other by cytoplasmic threads.

- Coenohium: When a colony has a definite number of cells with a definite shape and arrangement, it is called coenobium.

Coenobium may be:

- In motile form, cells are flagellated and whole coenobium can move by the organised beating action of flagella, e.g., Volvox, Pandorina, Eudorina etc. In Volvox the coenobium is a hollow sphere.

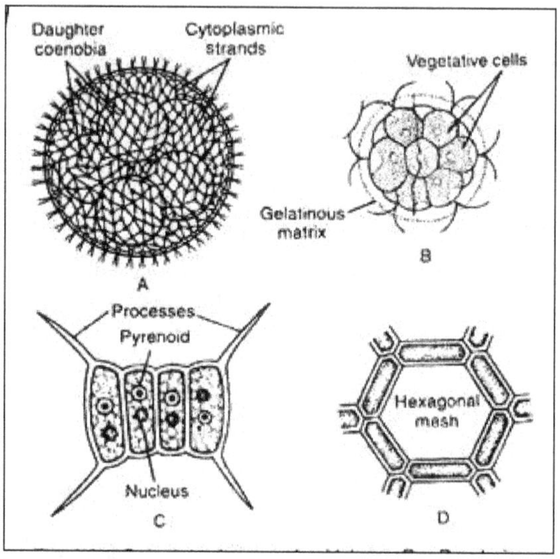

Figure: Colonial algae: A. Volvox, B. Pandorina, C. Scenedesmus, and D. Hydrodictyon.

- In non-motile form, the cells are without flagella, thereby the coenobium is non- motile, e.g., Scenedesmus, Hydro- dictyon.

Aggregated Form: Unlike coenobium the cells are aggregated irregularly showing a colonial mass of various size and shape.

It is of three types:

- Palmelloid.
- Dendroid.
- Rhizopodial.

Palmelloid

In this type the non- motile cells remain embedded in an amorphous gelatinous or mucilaginous matrix. Each and every cell of the organisation is independent and can perform all the functions as an individual. Chlamydomonas and Chromulina represent palmelloid as a temporary feature in their life cycle.

But in Tetraspora and Palmodictyon (Chlorophyceae), Gleochloris and Chlorosaccus (Xanthophyceae), Phaeocystis (Chrysophyceae) and Microcystis (Cyanophyceae), the palmelloid habit is a permanent feature.

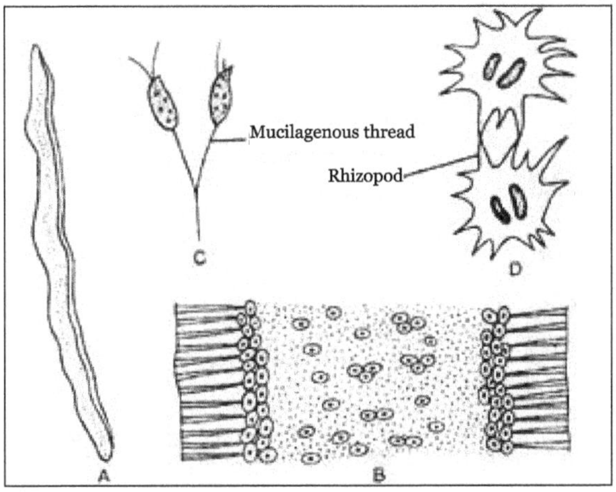

Figure: Aggregated form: A. Tetraspora, B. Tetraspora (portion of a colony in T.S.), C. Chrysodendron and D. Chrysidiastrum.

Dendroid

In this type the number, shape and size of the cell is variable. They look like microscopic trees (e.g., Prasinocladus, Ecballocystis, Chrysodendron etc.). A mucilaginous thread is present at the base of each cell, thus showing a sort of polarity.

Rhizopodial

In this type the cells are united through rhizopodia. E.g. Chrysidiastrum (Chrysophyceae).

Filamentous: The filamentous plant body is formed through repeated cell divisions in a single plane and in a single direction, where the cells remain firmly attached to each other — end to end forming a chain or a thread. The filaments may be unbranched or branched.

Unbranched Filament: It may be free-floating (e.g., Spirogyra) or attached to the substratum (e.g., Ulothrix, Oedogonium, etc.). The free-floating unbranched filaments are not differentiated into basal and apical ends. All the cells in the filament are alike. But the Unbranched filaments that remain attached to the substratum are differentiated into base and apex.

All the cells of the filament are similar except the basal attachment cell that is specially modified for the purpose called holdfast or rhizoidal cell. The cell is devoid of chloroplast and only performs the function of anchorage. So certain degree of division of labour among the cells of the filament is established as rest of the cells performs photosynthetic and reproductive functions.

- Branched Filament: It is formed when a filament occasionally starts division in a second plane. It is of two types:
    - Falsely branched.
    - Truly branched.
- Falsely Branched: The trichomes of blue greens may break either due to death or decay of the intercalary cells. The broken ends emerge out of the mucilaginous sheath in the form of a branch. They do not arise as lateral outgrowths, e.g., Scytonema.

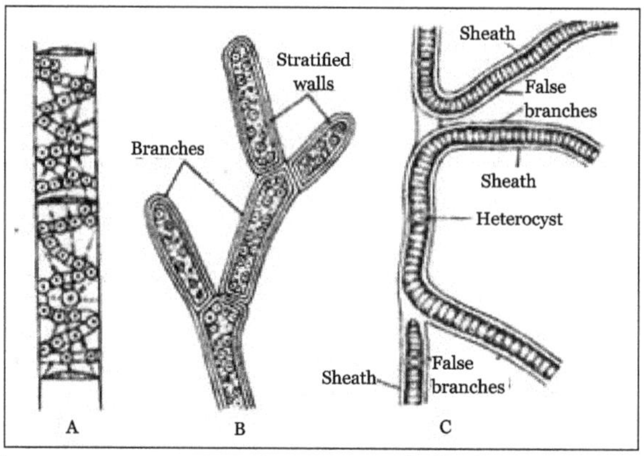

Figure: Filamentous type: A. Spirogyra, B. Cladophora and C. Scytonema.

Truly Branched

When a cell in the filament occasionally starts division in a second plane, true branch is formed. Thus true branches arise as lateral outgrowths of the main filament. True branches are of the following three types: Simple filament, Heterotrichous habit, and Pseudoparenchymatous habit.

Simple Filament: In this branching system the whole thallus remain attached to the substratum by a basal cell and the branches may arise from any cell of the filament except the basal cell, e.g., Cladophora.

Heterotrichous Habit: In this branching system the whole thallus is differentiated into prostrate and erect system. Both the prostrate and erect systems may be well-developed (e.g., Fritschiella, Ectocarpus). Progressive elimination of the prostrate system is observed in Draparnaldiopsis, Stigoclonium, and oi of the erect system as in Coleochaete.

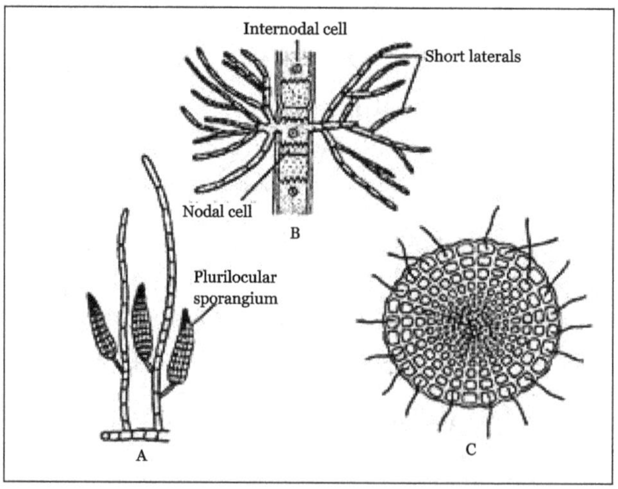

Figure: Branched filaments showing heterotrichous habit: A. Ectocarpus, B. Draparnaldiopsis, and C. Coleochaete.

Pseudoparenchymatous Habit: If one or more central or axial filaments together with their branches fuse to form a parenchymatous structure, it is called pseudoparenchymatous. Thallus again, if it is formed by the branches of a single filament it is known as uniaxial (e.g., Batrachospermum), or it may be multiaxial where more than one filament are involved.

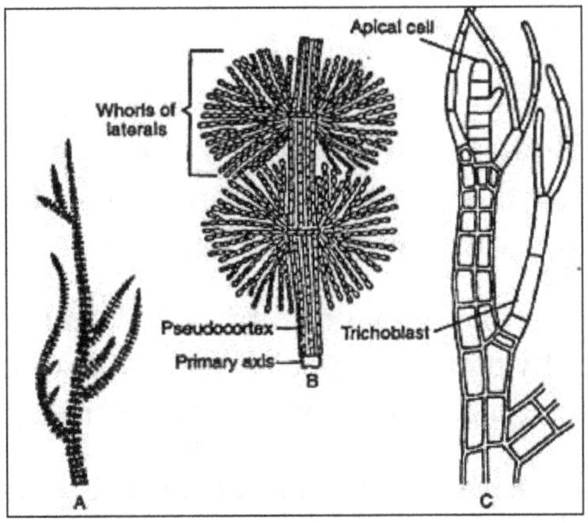

Figure: Pseudoparenchymatous habit: A. Batrachospermum (portion of the plant body), and C. Polysiphonia.

Siphonaceous Forms: In this form the thallus is aseptate and multinucleate i.e., a coenocyte. It may be simple branched (e.g., Vaucheria) or may be very elaborate with clear division of labour, differentiated into aerial and subterranean portions.

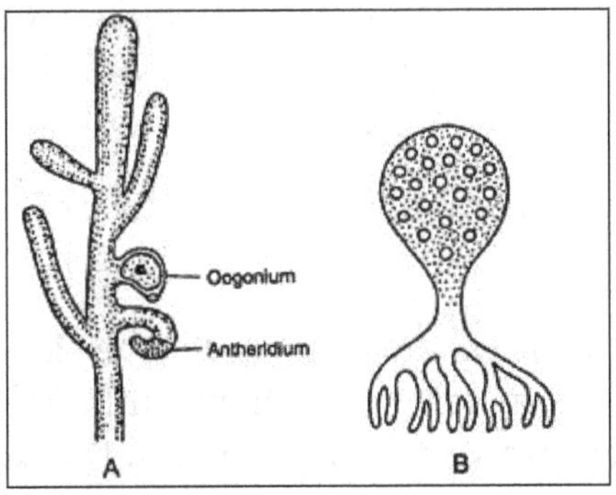

Figure: Siphonaceous algae: A. Vaucheria, and B. Botrydium.

Parenchymatous Forms: When the cells of a filament divide in multidirectional planes, it results the formation of a parenchymatous thallus and ultimately becoming foliose and flat (e.g., Ulva), tubular (e.g., Enteromorpha, Scytosiphon) or complex (e.g., Sargassum) structure.

Growth of the parenchymatous thalli may be diffused (when all the cells can divide), intercalary (when the dividing region remain in the intercalary position) e.g., Laminaria, trichothallic (growth by a specialised intercalary meristem at the base of a terminal hair) e.g., Porphyra or apical (when one or more well-defined apical cells divide to produce the remainder of the thallus), e.g., Fucus.

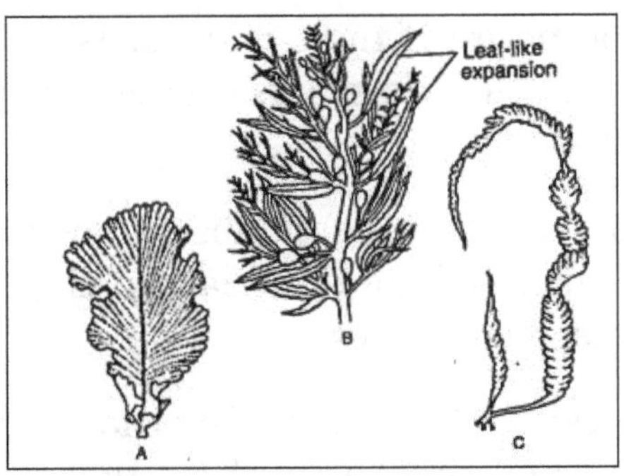

Figure: Parenchymatous algae: A. Ulva, B. Sargassum and C. Laminaria.

Origin and Evolution of Sex in Algae

Algae-like most of the other plants — reproduce by all the three means: vegetative, asexual, and sexual. The sexual reproduction is absent in the class Myxophyceae but they can reproduce by both vegetative and asexual means. In other groups the reproduction takes place by all the above three means, out of which asexual and sexual methods are very common.

Many plants multiply vegetatively, but they do not involve rejuvenation of the protoplasm. The asexual reproduction takes place by means of specialised motile or non-motile sex cells, the spores, which do not undergo fusion and, on germination, they give rise to new individuals.

Lastly, sexual reproduction involves the union of sex cells, the gametes, and the result of union of gametes is the zygote (2n), which on germination gives rise to new plant. The gametes are incapable of developing a new plant on germination.

Origin of Sex in Algae

The zoospores and gametes are developed during asexual and sexual reproduction, respectively. Both zoospores and gametes are morphologically alike except their size. The gametes are smaller in size than the zoospores. The origin of gamete is the starting point of the origin of sex.

The above fact can be interpreted by studying the life history of some algae like Chlamydomonas, Ulothrix etc.

Chlamydomonas debaryanum is the ideal member under the class Chlorophyceae. In this member the gametes and zoospores are alike in structure, shape and mode of development, but the difference lies in their size. The gametes are smaller in size than the zoospores. The above difference is visible due to the difference in the number of divisions in their maternal protoplasm.

During their formation, if the number of division is less, the unit protoplasts develop into zoospores. These zoospores have the sufficient amount of protoplasm to develop new plants on germination.

On the other hand, if the number of division is more, then the mother protoplast divides into more units and each unit develops into a structure like zoospore, but smaller in size and is incapable of germination into a new individual. Another member of the class Chlorophyceae U. zonata can produce three types of zoospores during asexual reproduction. These are:

- Quadriflagellate Macrozoospores: If there is no division of protoplast or the number of division is very less, single or a few zoospores are developed. This zoospore on germination develops into healthy plant.
- Quadriflagellate Microzoospores: If the number of division of protoplast is more, more number of zoospores is formed and, on germination, they develop new plants, weaker than the plants developed by macrozoospores.
- Biflagellate Microzoospores: If the number of division of protoplast is still more, the sporangium forms large number of unit of protoplasts, those form biflagellate microzoospores.

These microzoospores, on germination, develop into plants, those are still weaker than the above two cases.

The microzoospores are alike in structure and show similar mode of development like gametes. During unfavourable condition, the microzoospores fail to liberate from the sporangium and undergo more divisions and thus form more number of smaller units.

These smaller units behave as gametes. These gametes undergo fusion to form zygote. The zygote takes rest and during favourable condition germinates into a new plant, which bears asexual spores again.

Based on the above discussion it has been postulated that during unfavourable condition, sexuality in algae have originated as a result of accidental fusion of very small microzoospore- like units which are incapable of developing new individuals.

The above view is also supported by the fact that in lower group of plants sexual reproduction takes place during unfavourable condition to overcome the situation. According to the "starvation theory" of Cholnoky, the sexuality is originated in algae due to attraction between two nutritionally deficient cells.

Evolution of Sex in Algae

The evolution of sex takes place by a different process from simple isogamy to complex heterothallic oogamy through physiological and morphological anisogamy.

Isogamy

In the primitive and simplest form like Chlamydomonas debaryanum, Cladophora etc., both the fusing gametes are morphologically and physiologically identical, thus they cannot be differentiated into male and female gametes. The gametes are called isogametes and the process is called isogamy.

Physiological Anisogamy

In some algae, the gametes are morphologically alike, but differ in their physiological behaviour as:

- In Ulothrix, the gametes thus produced are morphologically identical, but the fusion takes place between gametes originating from the different filaments indicate the difference in their physiological characteristics and can be designated as + and – gametes.
- In Spirogyra, the gametes are non- motile and identical in shape and size; those develop singly within the cell. At the time of conjugation the two filaments come very close to each other and some of the cells are connected by conjugation tube.

Out of the two fusing gametes one becomes passive and remains within the cell and behaves as female gamete. On the other hand, other gamete though non-motile becomes active and passes to the female through conjugation tube and behaves as male gamete.

Thus, though the gametes are morphologically identical, they show difference in their behaviour i.e., the physiological anisogamy.

Anisogamy

In Ectocarpus, Pandorina, Clodium and Chlamydomonas braunii, the anisogamy is directly visible, here both the gametes are ciliated i.e., motile, but unequal in size. The gametes are called anisogametes. The smaller one is called micro- gamete which behaves as male and the larger one is called macrogamete which behaves as female. The micro- and macro- gamete are produced within the micro- and macrogametangium, respectively.

Oogamy

This type of sexual union is visible in Chlamydomonas oogamum, C. coccifera etc. Here male and female gametes are produced within antheridia and oogonia, respectively. The smaller one is active and called male gamete or antherozoid but the relatively larger one is inactive and called female gamete or egg.

Usually single egg is formed within oogonium except in Fucus and Sphaeroplea. Many male gametes are formed within the antheridium.

In Phaeophyceae, both male and female gametes are discharged from the antheridium and oogonium, respectively, and their union occurs in water. This type is called primitive oogamy.

In Oedogonium, the male gametes i.e., antherozoids, are smaller, flagellated and develop in pair within unicellular antheridium, but the female gamete i.e., egg, develops singly within oogonium. The fertilisation takes place within oogonium.

In Chara, the sex organs are further specialised. The round male sex organ is the globule containing huge number of antherozoids and the more or less oval, much protected structure is called nucule containing only one egg. The protection of egg and zygote is much more, indicating an advanced characteristic.

In Fucus, separate male, female and mixed conceptacles are formed on receptacles. Out of eight eggs developed in oogonium, seven degenerate. Till now all the species are homothallic.

The evolution of sex reaches its climax in the heterothallic species of Rhodophyceae. Spermatia, the male gametes, are non-motile and developed singly in spermatangium, those are carried by water current to the trichogyne, the receptive region of the female sex organ — the carpogonium.

In Polysiphonia and Oedogonium, out of four tetraspores or zoospores developed (by meiosis) from tetrasporangia or directly from zygote, two produce female plants and other two male plants.

Fungi

Fungus is a eukaryote that digests food externally and absorbs nutrients directly through its cell walls. Most fungi reproduce by spores and have a body (thallus) composed of microscopic tubular cells called hyphae. Fungi are heterotrophs and, like animals, obtain their carbon and energy from other organisms. Some fungi obtain their nutrients from a living host (plant or animal) and are called biotrophs; others obtain their nutrients from dead plants or animals and are called saprotrophs (saprophytes, saprobes). Some fungi infect a living host, but kill host cells in order to obtain their nutrients; these are called necrotrophs.

Fungi were once considered to be primitive members of the plant kingdom, just slightly more advanced than bacteria. We now know that fungi are not primitive at all. In fact, recent taxonomic treatments such as the Tree of Life Project show that fungi and animals both belong to the group Opisthokonta. Fungi may not be our next of kin, but they are more closely related to animals than they are to plants. We also recognize that organisms traditionally studied as "fungi" belong to three very different unrelated groups: the true fungi in Kingdom Fungi (Eumycota), the Oomycetes, and the slime molds.

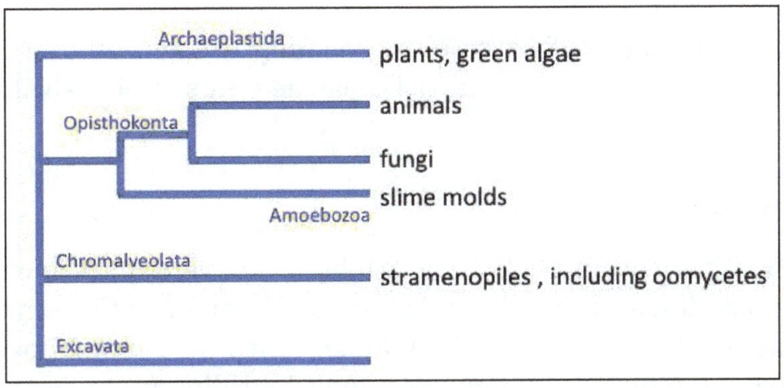

Fungi are an ancient group—not as old as bacteria, which fossil evidence suggests may be 3. 5 billion years old—but the earliest fungal fossils are from the Ordovician, 460 to 455 million years old. Based on fossil evidence, the earliest vascular land plants didn't appear until approximately 425 million years ago, and some scientists believe that fungi may have played an essential role in the colonization of land by these early plants. Mushrooms exquisitely preserved in amber from the Late Cretaceous (94 million years ago) tell us that there were mushroom-forming fungi remarkably similar to those that exist today when dinosaurs were roaming the planet. However, the fungal fossil record is incomplete and provides only a minimum time estimate for when different groups of fungi evolved. Molecular data suggest that fungi are much older than indicated by the fossil record, and may have arisen more than one billion years ago.

Types of Fungi

No one knows for sure how many species of fungi there are on our planet at this point in time, but what is known is that at least 99,000 species of fungi have been described, and new species are described at the rate of approximately 1200 per year. A conservative estimate of the total number of fungal species thought to exist is 1.5 million. To come up with this figure, Hawksworth estimated the known numbers of plant and fungal species from countries in which both plants and fungi have been well-studied—Great Britain and Ireland, in this case—and determined there were six fungal species for every native plant species. The total number of plant species worldwide is approximately 250,000, and if the ratio of fungi to plants in Great Britain is typical of what occurs elsewhere, there should be at least 1.5 million species of fungi.

If 1.5 million fungal species is a reasonable estimate, the vast majority of all extant fungi are yet to be named. Assuming a relatively constant rate at which new species are described, it will take more than 1100 years to catalog and describe all remaining fungi. However, many of these fungi are likely to become extinct before they are ever discovered given current rates of habitat and host loss. For example, up to 2% of tropical forests are destroyed globally each year. These habitats are exceedingly rich in fungal species. For example, 15-25% of fungi collected in short-term studies in the tropics are new species. Callan and Carris estimated that an 110,000 ha neotropical forest, such as in Costa Rica, could contain over 81,000 different species of plant parasitic fungi—almost as many as all the known species of fungi! Consider that this estimate was based only on plant parasitic fungi, and did not take into account other ecological groups of fungi such as saprotrophs.

Functions of Fungi

Fungi are involved in a wide range of activities—some fungi are decomposers, parasites or pathogens of other organisms and others are beneficial partners in symbiosis with animals, plants or algae.

Fungi Associated with Animals

Fungi have the ability to grow on and in both invertebrate and vertebrate animals. Many fungi can attack insects and nematodes, for example, and may play an important role in keeping populations of these animals under control. Insect-attacking fungi, called "entomopathogens," include a wide range of fungi in phyla Ascomycota, Zygomycota and Chytridiomycota. Some of the best-known and most spectacular entomopathogens belong in the Ascomycota genus Ophiocordyceps and related genera. These fungi infect and consume insects such as caterpillars and ants, and then form conspicuous stromata that emerge from their victim's body in a most dramatic manner. These fungi can also alter the insect's behavior. "Zombie-ant" fungi from Brazil infect insect brains, directing the victim to climb up plants and bite into the plant tissue in a "death grip". Paradoxically, humans have been using one of these entomopathogens, Ophiocordyceps sinensis, for thousands of years to treat a wide range of ailments. This fungus is an important component of traditional Asian medicine and is commonly called "winter worm, summer grass."

Entomopathogens such as Beauveria bassiana are so effective in killing insects that they are used as biological control agents for insect pests. Colony collapse disorder of honeybees has been associated with co-infection by a virus and a microsporidian fungus, Nosema ceranae. One group of fungi called Entomophthorales ("insect killers") includes a number of highly specialized entomopathogens. A common example is Entomophthora musae, which is often observed forming a ring of white spores discharged around the body of a parasitized fly on panes of glass.

Some fungi are specialized parasites of nematodes, rotifers, and other microscopic animals in the soil. A common nematode predator is Arthrobotrys oligospora, a fungus that has evolved sticky networks of hyphae for trapping nematodes. Once the nematode is immobilized, the fungus invades and consumes its body.

Fortunately, there are relatively few fungal pathogens of vertebrates—only 200-300 species—but some of these fungi can have devastating impacts. Consider the well-publicized frog killer, Batrachochytrium dendrobatidis, a member of phylum Chytridiomycota. This fungus wasn't even known to scientists until 1996, when it was discovered associated with frogs that had died from a mysterious skin disease at the Smithsonian National Zoological Park in Washington, D. C. The fungus doesn't invade the frog's body, but it is lethal, possibly because it disrupts electrolyte balance leading to cardiac arrest—infected frogs appear to die of a heart attack! The frog chytrid is implicated in the widespread decline of frog populations around the world. Fortunately, this is the only chytrid known to parasitize a vertebrate animal and it appears to infect only amphibians.

Another devastating parasite of animals is Geomyces destructans, a cold-loving fungus that causes 'white-nose syndrome' in bats. This fungus colonizes the skin on the muzzles, ears and wing membranes of some types of bats, and infected bats exhibit unusual behavior. The bat fungus is associated with declines in bat populations in the northeastern U. S. and has many wildlife biologists concerned. As of 2011, white-nose syndrome had been confirmed in 16 states and four Canadian provinces.

In humans, there are several different types of fungal infections, or "mycoses." The most common are caused by dermatophytes, fungi that colonize dead keratinized tissue including skin, finger-, and toenails. Dermatophytes cause superficial infections such as 'ringworm' that are unsightly and difficult to treat, but rarely serious. Some fungi are members of the resident microflora in healthy people, but become pathogenic in people with predisposing conditions. For example, Candida species cause annoying yeast infections in the mucosal tissues of many healthy people, but can also cause diseases collectively called candidiasis in babies and immunocompromised individuals. Another group of fungi are inhaled as spores and initiate infection through the lungs. These fungi include Coccidioides immitis (coccidioidomycosis, commonly known as valley fever), and Histoplasma capsulatum (histoplasmosis). Opportunistic fungal pathogens are normally not associated with humans and other animals, but can cause serious infections in weakened or healthy individuals when inhaled or implanted in wounds. Aspergillus fumigatus, one of the most important of these opportunists, produces small, airborne spores that are frequently inhaled; in some individuals the fungus starts growing invasively, causing a disease known as aspergillosis, especially in immunocompromised individuals.

A remarkable discovery was that Pneumocystis carinii, the organism causing pneumonia-like symptoms in immunocompromised patients, is a fungus and not a protozoan as had been thought for decades. Why was this pathogen classified as a protozoan? It does not respond to the common drugs used to treat fungal infections, but does respond to anti-protozoan drugs. This unusual fungus emerged as one of the leading causes of death in AIDS patients in the late twentieth century.

Fungi and Plants

The association of fungi and plants is ancient and involves many different fungi. Fungi are an important group of plant pathogens—most plant diseases are caused by fungi—but fewer than 10% of all known fungi can colonize living plants. Plant pathogenic fungi represent a relatively small subset of those fungi that are associated with plants. Most fungi are decomposers, utilizing the remains of plants and other organisms as their food source.

Most fungi are associated with plants as saprotrophs and decomposers. These fungi break down organic matter of all kinds, including wood and other types of plant material. Wood is composed primarily of cellulose, hemicellulose, and lignin. Lignin is a complex polymer that is highly resistant to degradation, and it encrusts the more readily degradable cellulose and hemicellulose. Fungi are among the few organisms that can effectively break down wood, and fall into two main types—brown and white rot fungi. Brown rot fungi selectively degrade the cellulose and hemicellulose in wood, leaving behind the more recalcitrant lignin. The decayed wood is brown in color and tends to form cubical cracks due to the brittle nature of the remaining lignin. Only ~ 10% of the wood decay fungi cause brown rot, and most of these fungi (80%) occur on conifer wood. Brown rot residues make up 'humus' in temperate forest soils and are important for mycorrhizal formation, moisture retention, and for sequestering carbon. Brown rot residues are highly resistant to decomposition and can remain in the soil for up to 300 years. White rot fungi are more common than brown rot fungi; these fungi degrade cellulose, hemicellulose, and lignin at approximately equal rates. The decayed wood is pale in color, light in weight, and has a stringy texture. White rot fungi are the only organisms that can completely degrade lignin. Lignin is one of the most abundant organic polymers, accounting for 30% of the organic carbon on the planet—only cellulose is more abundant.

An important group of fungi associated with plants is mycorrhizal fungi. Mycorrhiza means 'fungus root', and it refers to a mutually beneficial association (a type of symbiosis) between fungi and plant roots. There are seven major types of mycorrhizal associations, the most common of which is the arbuscular mycorrhizae, involving members of phylum Glomeromycota associated with roots of most major groups of plants.

Another common type of association is ectomycorrhizae formed between forest trees and members of phyla Basidiomycota and Ascomycota. In this association, the fungus forms hyphae around host root cortical cells—the "Hartig net"— and a sheath of hyphae around the host roots called a "mantle". Many of the ectomycorrhizal fungi are mushroom-forming species including highly prized edibles such as chanterelles (Cantharellus cibarius and related species), boletes (Boletus edulis and related species), and matsutake (Tricholoma magnivelare). A valuable group of ectomycorrhizal fungi are truffles, members of phylum Ascomycota that form underground fruiting bodies.

Lichens are examples of a symbiotic association involving a fungus and green algae or less frequently Cyanobacteria. The lichen thallus is composed mostly of fungal hyphae, usually with the alga or cyanobacterium confined to discrete areas of the thallus. In lichens, reproductive structures of the fungus are often conspicuous, for example disc- or cup-like structures called apothecia. The fungus obtains carbohydrates produced by photosynthesis from the algae or cyanobacteria, and in return provides its partner(s) with protection from desiccation and ultraviolet light. Lichens grow in a wide range of habitats on nearly every continent. Think about an inhospitable place, and there's probably lichen that grows there—on bare rocks, sidewalks, grave stones, the exoskeletons of some insects, and even on cars that remain for a long time in one place.

Some fungi are hidden inside their plant hosts; these are endophytes, defined by their presence inside asymptomatic plants. All plants in natural ecosystems probably have some type of symbiotic association with endophytic fungi. Endophytic fungi have been shown to confer stress tolerance to their host plant, for example, to disease, herbivory, drought, heat, salt and metals. The clavicipitaceous endophytes in the genus Neotyphodium (phylum Ascomycota) are among the best studied. These fungi produce alkaloid compounds that protect the grass host from insects that would otherwise feed on them; endophyte-infected turf grass seed is sold commercially for seeding lawns and other types of grassy recreational areas. Unfortunately, livestock such as sheep, cattle, llamas and horses also are negatively affected by toxins produced by endophytes when they eat infected grass. 'Ryegrass staggers' occurs when animals graze on perennial ryegrass (Lolium perenne) that is colonized by Neotyphodium lolii. Afflicted animals develop symptoms including tremors and jerky or uncoordinated movements.

Role of Fungi as Plant Pathogens

There are thousands of species of plant pathogenic fungi that collectively are responsible for 70% of all known plant diseases. Plant pathogenic fungi are parasites, but not all plant parasitic fungi are pathogens. What is the difference between a parasite and a pathogen? Plant parasitic fungi obtain nutrients from a living plant host, but the plant host doesn't necessarily exhibit any symptoms.

Biotrophic fungal pathogens obtain nutrients from living host tissues, often via specialized cells called haustoria that form inside host cells. Necrotrophic pathogens obtain nutrients from dead host tissue, which they kill through the production of toxins or enzymes. Most biotrophic fungi have fairly narrow host ranges—they are specialized on a limited number of plant hosts. Necrotrophic fungi can be either generalist, growing on a wide range of host species, or specialized on a restricted range of hosts. Some plant pathogenic fungi change the way that their hosts grow, either by affecting the level of growth regulators produced by the plant, or by producing growth regulators themselves. Examples of changes in plant growth caused by plant pathogenic fungi include cankers, galls, witches' broom, leaf curl and stunting.

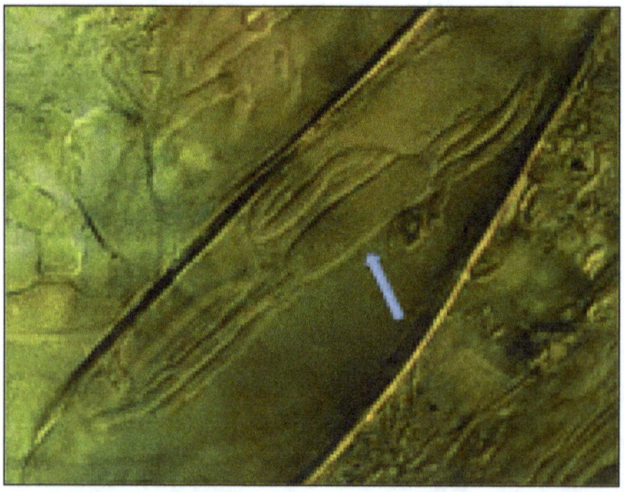

We can further divide plant pathogenic fungi by the stage of the plant host that is attacked, for example, seeds, seedlings, or adult plants, and by what part of the plant is affected—roots, leaves, shoots, stems, woody tissues, fruits or flowers. A group of fungi including species of Fusarium, Rhizoctonia and Sclerotium cause seed rot and infect plants at the seedling stage. These pathogens can attack a wide range of plants. Often, seedling pathogens cause damping-off symptoms because they occur in wet soils.

Many of the same fungi that kill seedlings can also infect the roots of mature plants and cause root and crown rot diseases. Infection often occurs through wounds, and results in lesions or death of part or all of the root system and crown. Some common root rots of trees are caused by members of phylum Basidiomycota in the genera Armillaria and *Heterobasidion*. Armillaria spp. produce shoe-string-like bundles of hyphae called rhizomorphs that allow the fungus to grow from one tree to another. Species of Heterobasidion survive as saprotrophs in dead tree stumps and roots, but can also infect living hosts through root contact. These fungi cause decay in the roots and crown; infected trees become weakened and die, or may blow over in high winds. Wood rot fungi, most of which are also members of Basidiomycota, infect trees through wounds, branch stubs and roots, and decay the inner heartwood of living trees. Extensive decay weakens the tree, and reduces the quality of wood in trees harvested for timber.

Vascular wilt pathogens kill their host by infecting through the roots or through wounds and growing into the xylem, where they produce small spores that get carried upward until they are trapped at the perforated ends of the xylem vessels. The spores germinate and grow through the pores. The fungus is transported throughout the plant in this manner. The first symptom of vascular wilt is a loss of turgidity in the plant leaves, often on one side of the plant or a single branch. If the stems of infected plants are cut open, vascular discoloration is evident. Among the important vascular wilt fungi are Fusarium oxysporum, Verticillium albo-atrum and V. dahliae.

One of the most famous vascular wilts is Panama disease of bananas, caused by Fusarium oxysporum forma specialis (f. sp.) cubense. Most bananas that were being grown for export were a single cultivar, 'Gros Michel', which turned out to be highly susceptible to Panama disease. There is no effective method for controlling Panama disease and it rapidly spread throughout banana plantations around the world. The banana industry was saved by the discovery of the cultivar 'Cavendish' that is resistant to the strain of Panama disease that killed 'Gros Michel'. 'Cavendish' is now the banana that Americans and Europeans consume, but a new strain of the Panama disease pathogen began killing 'Cavendish' in Malaysia in 1985 and scientists are concerned that this strain will begin to spread.

Leaf spot pathogens infect through natural plant openings such as stomates or by penetrating directly through the host cuticle and epidermal cell wall. In order to penetrate directly, fungi produce hydrolytic enzymes—cutinases, cellulases, pectinases and proteases—for breaking down the host tissue. Alternatively, some fungi form specialized structures called appressoria (sing. appressorium) at the end of germ tubes. Turgor pressure builds up in the appressorium, and in combination with an infection peg, mechanical force is exerted to breach the host cell walls. Once inside the plant leaf, the fungus must obtain nutrients from the cells, and this is often accomplished by killing host cells (necrotrophs). Death of host cells is evident as an area of dead cells called a lesion.

Many leaf-spotting fungi produce toxins that kill host cells and this often produces a lesion surrounded by a yellow halo. If enough of the leaf surface is killed, or if the infected leaves drop prematurely, the plant's ability to produce photosynthates is severely impaired.

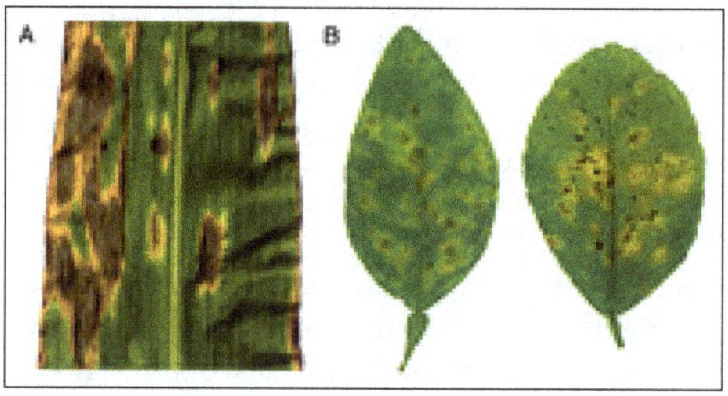

Returning to bananas, another devastating disease of this host is black leaf streak, or black Sigatoka, caused by Mycosphaerella fijiensis. Unlike the root-infecting pathogen that causes Panama disease, the black Sigatoka pathogen can be controlled by applications of a protective fungicide to banana leaves. Effective control of black Sigatoka requires multiple fungicide applications and control of this disease accounts for up to 25% of the total production cost for bananas.

American chestnut trees were once a prominent hardwood tree in the eastern U. S., but have largely been eliminated by the chestnut blight pathogen, Cryphonectria parasitica, an example of a canker-causing fungal pathogen. Cankers develop when the pathogen kills the phloem and vascular cambium in a woody host. If the canker encircles the trunk or branch of a tree, that plant part will die. The canker-causing fungus can often be identified based on the fruiting bodies that form in the canker. In contrast to cankers, galls result from abnormal growth of a plant, usually due to an increase in cell size and cell division. Although galls are often associated with insect pests, some fungal pathogens induce galls; two common examples are the black knot pathogen, Apiosporina morbosa on Prunus spp., and species of Gymnosporangium, which induce the formation of galls on their coniferous hosts.

Gymnosporangium is a type of rust fungus. Rust fungi are biotrophic pathogens—they infect, grow, and sporulate in living plant tissue. Even though biotrophs require living host tissue for their growth and reproduction, they can be devastating pathogens by reducing the photosynthetic surface and increasing water loss in the host plant. Rust fungi attack a wide range of plants, and often require two, unrelated hosts in order to complete their life cycles. Rust fungi are so-named because of the abundant orange spores that are formed on plants that are infected by these fungi; infected plants often look as though they are rusting.

One historically important rust fungus is black stem rust of wheat, a disease that was well known to the ancient Romans. Black stem rust, caused by Puccinia graminis f. sp. tritici, infects wheat and barberry (Berberis species). Since the barberry host is required for the pathogen to complete its life cycle, early control measures in the United States and Canada were aimed at eliminating this host, not the economically important host, wheat.

Morphological Characteristics of Fungi: Mycelium and Hyphae

Now let's take a closer look at fungi and the types of structures that they form. A key characteristic of fungi that has contributed to their successful exploitation of diverse ecological niches is the formation of a filamentous thallus called the mycelium. A mycelium is composed of branching, microscopic tubular cells called hyphae that grow through and across substrates or food sources, secreting enzymes that break down complex substrates into simple compounds that can be absorbed back through the cell wall. The fungal cell wall in the Kingdom Fungi is composed of chitin and glucans (in Ascomycota, Basidiomycota and Chytridiomycota) as well as chitosan and other components (in Zygomycota).

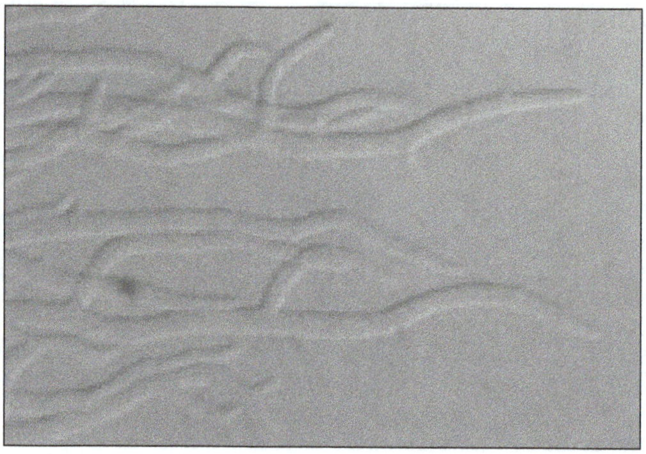

Hyphae can have cross walls called septa, or lack cross walls (nonseptate; aseptate; coenocytic). The type of hyphae—septate or aseptate—is characteristic of specific groups of fungi. In fungi that form septate hyphae, there are perforations at the septa, called septal pores, which allow the movement of cytoplasm and organelles from one compartment to the next. The type and complexity of the septal pore is characteristic of specific groups of fungi.

Hyphae elongate almost exclusively at the tips, growing outwards from the point of establishment. As a result of apical growth, hyphae are relatively uniform in diameter, and mycelium that grows in an unimpeded manner forms a circular colony on solid substrates that support fungal growth; agar, a gelatinous material derived from seaweed, amended with different types of nutrients is commonly used to grow fungi in culture.

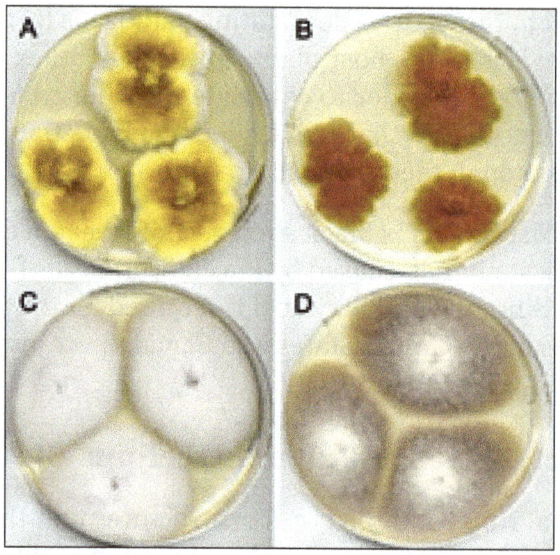

Some fungi grow exclusively or mostly as yeasts, defined as single-celled fungi that reproduce by budding or fission. In contrast to apical growth that is characteristic of hyphae, yeasts exhibit wall growth over the entire cell surface, often resulting in a nearly spherical cell. There are also fungi that can switch between mycelial growth and yeast-like growth, dependent upon the environmental conditions. The ability to grow in different forms is called dimorphism, and is exhibited by some members of phyla Ascomycota, Basidiomycota and Zygomycota.

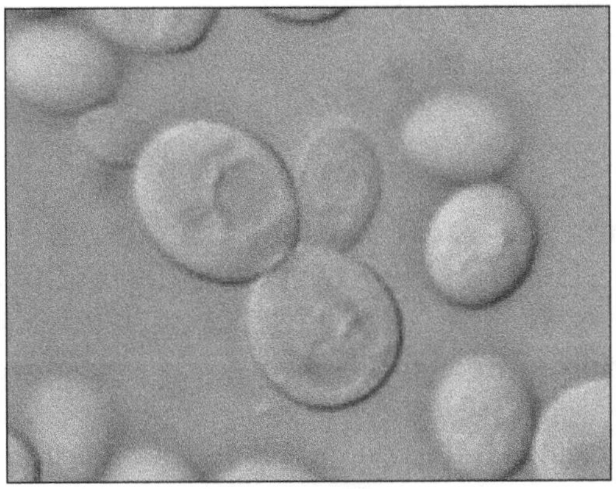

Inside the Fungal Cell

Most of the organelles present in fungal cells are similar to those of other eukaryotes. Fungal nuclei are usually small (< 2 μm diameter), and can compress and/or stretch to move through septal pores and into developing spores. Fungi have been found to possess between 6 and 21 chromosomes coding for 6,000 to nearly 18,000 genes. Genome sizes range from 8. 5 mega base pairs (Mb) to just over 400 Mb in filamentous fungi, making fungal genomes among the smallest of eukaryotic organisms on average—approximately 1% the size of mammalian genomes and only 1.3 times the size of the largest known bacterial genome. Many fungi (Ascomycota) have a life cycle that is predominantly haploid, while others (Basidiomycota) have a long dikaryotic phase.

Fungal Reproduction

Fungi frequently reproduce by the formation of spores. A spore is a survival or dispersal unit, consisting of one or a few cells, that is capable of germinating to produce a new hypha. Unlike plant seeds, fungal spores lack an embryo, but contain food reserves needed for germination. Many fungi produce more than one type of spore as part of their life cycles. Fungal spores may be formed via an asexual process involving only mitosis (mitospores), or via a sexual process involving meiosis (meiospores). The manner in which meiospores are formed reflects the evolutionary history and thus the classification for the major groups (phyla) of fungi.

Many fungi produce spores inside or upon a fruiting body. Many people are familiar with the mushroom, a type of fruiting body produced by some Basidiomycota. You may recognize other fungal fruiting bodies such as puffballs, or shelf fungi. These are examples of large, conspicuous fruiting bodies, but there is an even greater diversity of microscopic fruiting bodies produced by various fungi. What all fruiting bodies have in common is that they produce spores and provide a mechanism for dispersing those spores.

Teleomorph and Anamorph

Many fungi are able to reproduce by both sexual and asexual processes. Sexual and asexual reproduction may require different sets of conditions (e. g., nutrients, temperature, light, moisture). In some fungi, two sexually compatible strains must conjugate (mate) in order for sexual reproduction to occur. The terms 'anamorph' and 'teleomorph' are used to convey the asexual and sexual reproduction morphological types, respectively, in a particular fungus. The concept of anamorph and teleomorph is a confusing one for many students, as we are not accustomed to thinking about organisms with such reproductive flexibility.

Meiospores

Examples of meiospores—spores that are the products of meiosis—include ascospores and basidiospores. Ascospores are formed inside a sac-like structure called an ascus. An ascus starts out as a sack of cytoplasm and nuclei, and by a process called "free cell formation" a cell wall forms de novo around each nucleus and surrounding cytoplasm to form ascospores (typically eight per ascus). Ascospores vary in size, shape, color, septation, and ornamentation among taxa. Basidiospores are formed on a basidium and are typically one-celled with one or two haploid nuclei. Basidiospores vary in size, color and ornamentation depending upon the taxonomic group. More information on dispersal of ascospores and basidiospores can be found below.

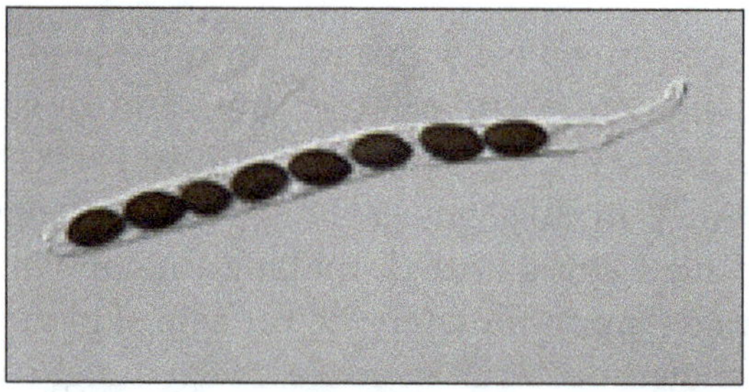

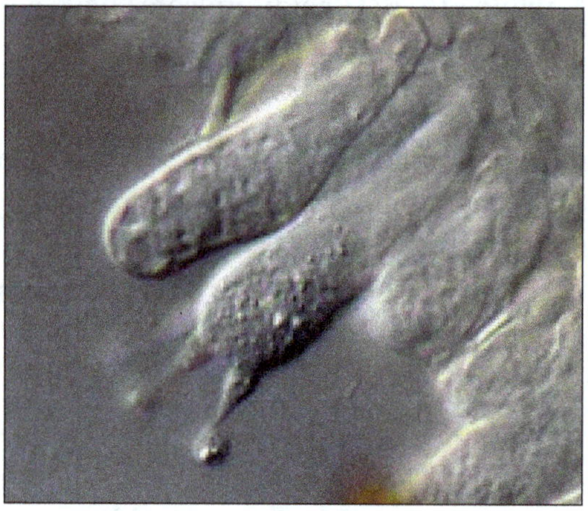

Mitospores

Examples of mitospores are conidia (sing. conidium), sporangiospores, and zoospores, formed by members of the phyla Ascomycota, Zygomycota, and Chytridiomycota, respectively. Another type of asexual propagule produced by fungi in several different phyla is the chlamydospore.

Conidia

Conidia are formed from a modified hypha or a differentiated conidiogenous cell of the fungus. Conidiogenous cells can be formed singly on hyphae, on the surface of aggregated hyphal structures, or within different types of fruiting bodies. Fruiting bodies inside which conidia are formed are pycnidia and acervuli. Sporodochia and synnemata are examples of fruiting bodies on which conidia are formed. Conidia are produced primarily by Ascomycota, although some Basidiomycota are capable of producing them as well.

Sporangiospores

Sporangiospores are asexual propagules formed inside a globose or cylindrical sporangium by a process involving cleavage of the cytoplasm. Sporangiospores are thin-walled, one-celled, hyaline or pale-colored, and are usually globose or ellipsoid in shape. One to 50,000 sporangiospores may be formed in a single sporangium. When mature, sporangiospores are released by breakdown of the sporangial wall, or the entire sporangium may be dispersed as a unit. Sporangiospores are produced by fungi in phyla Chytridiomycota and Zygomycota, as well fungal-like Oomycetes.

Zoospores

A zoospore is a microscopic, motile propagule, approx. 2 to 14 µm long and 2 to 6 µm in diameter that lacks a cell wall and is characterized by having one or more flagella. Flagella are ~ 0.25 µm in diameter and up to 50 µm long. Zoospores are produced by one group of true Fungi (Chytridiomycota), and by fungal-like organisms in Kingdom Straminipila and some slime molds. Two types of flagella are known—the whiplash flagellum, which is directed backward, and the tinsel flagellum, which is directed forward. The tinsel flagellum is only present in members of Kingdom Straminipila and does not occur in true fungi. The length of time zoospores are able to swim is determined by their endogenous energy reserves—zoospores cannot obtain food from external sources—and environmental conditions. Zoospores may exhibit chemotaxis—movement in response to a chemical gradient, e. g., root exudates. At the end of its motile phase, the zoospore undergoes a process called encystment in which it either sheds or retracts the flagella and produces a cell wall. The encysted zoospore, called a cyst, may germinate directly by the formation of a germ tube, or indirectly by the emergence of another zoospore.

Zoospores are formed inside a sac-like structure called a zoosporangium by a process involving mitosis and cytoplasmic cleavage—similar to the formation of sporangiospores in sporangia. Depending upon the taxonomic group, zoospores emerge from the zoosporangium through breakdown of the zoosporangial wall, through a preformed opening in the wall covered with a cap called an operculum that flips back, or by a gelatinous plug that dissolves.

Chlamydospores

Chamydospores are survival propagules formed from an existing hyphal cell or a conidium that develops a thickened wall and cytoplasm packed with lipid reserves. The thickened cell walls may be pigmented or hyaline, and chlamydospores develop singly or in clusters, depending upon the fungus. Chlamydospores are passively dispersed, in most instances when the mycelium breaks down. Chlamydospores are formed by many different groups of fungi and are often found in aging cultures.

Sclerotia

Sclerotia (sing. sclerotium) are compact aggregations of hyphae differentiated into an outer, pigmented rind, and an inner mass of hyaline cells called a medulla. Sclerotia contain food reserves, and are a type of survival propagule produced by a number of fungi in phyla Ascomycota and Basidiomycota; in some fungi, such as Rhizoctonia solani, they are the only type of propagule produced, whereas in fungi such as Claviceps purpurea and Sclerotinia sclerotiorum, they are overwintering structures that can germinate directly, or give rise to structures in which the meiospores are formed.

Kingdom Fungi

The characteristics and diversity of the major phyla of true Fungi will be briefly described. Selected representatives of the different phyla are introduced and, in many instances, illustrated. A generalized life cycle also is presented for each phylum that illustrates when plasmogamy (cell fusion), karyogamy (nuclear fusion) and meiosis occur relative to each other, and the types of structures involved in these events.

Phylum Ascomycota is the largest group of fungi, with approximately 33,000 described species in three subphyla—Taphrinomycotina, Saccharomycotina, and Pezizomycotina. Members of this phylum reproduce sexually or meiotically via production of ascospores inside a sac-like structure called an ascus.

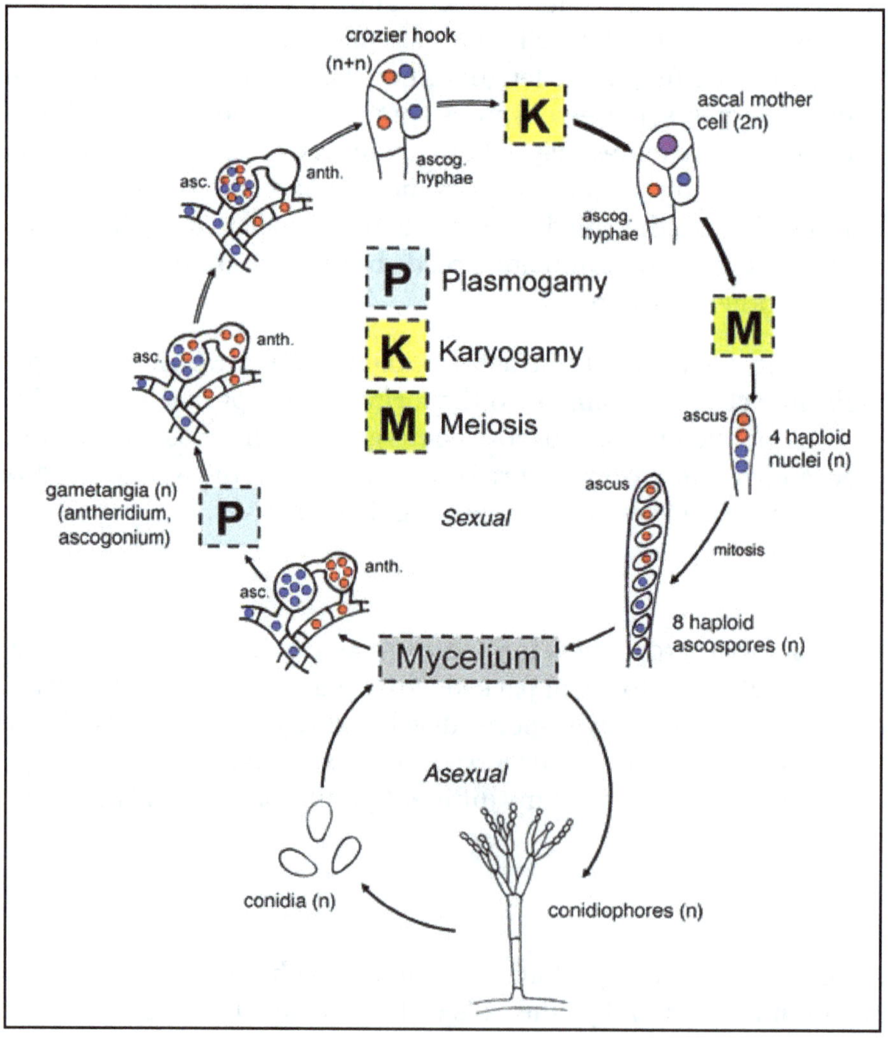

Many species of Ascomycota also (or exclusively) produce spores through an asexual or mitotic process; these spores, called conidia, exhibit a wide range of size, shape, color and septation among the different fungi in which they are formed. Conidia and ascospores are usually produced at different times of year, if ascospores are formed in the lifecycle. The existence of many Ascomycota having sexual and asexual states that are separated in time and space has long confused those new to mycology and plant pathology. The asexual states of Ascomycota are especially important to the plant pathologist because they are more commonly encountered than the sexual state, and must be identified for control, quarantine, or other purposes. Fungi that reproduce only via asexual means have been given various designations including deuteromycetes, fungi imperfecti, mitosporic fungi, conidial fungi, and anamorphic fungi.

Subphylum Taphrinomycotina includes fungi that, with one known exception, do not form fruiting bodies—as examples, the fission yeast Schizosaccharomyces, plant parasites in the genera Protomyces and Taphrina (peach leaf curl), and Pneumocystis, a yeast-like fungus.

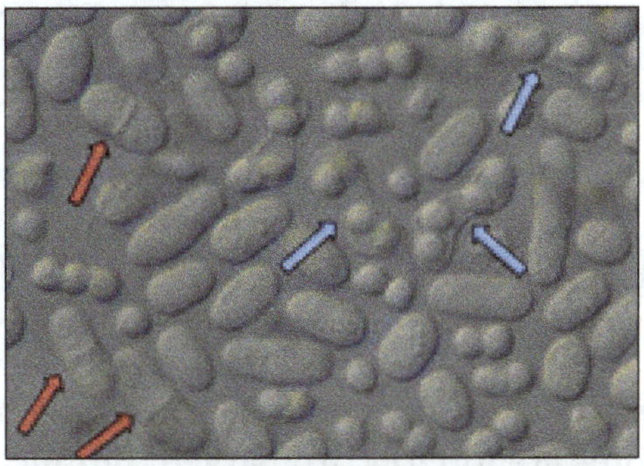

Subphylum Saccharomycotina contains approximately 1500 species of yeasts, most of which live as saprotrophs in association with plants and animals, but also including a small number of plant and animal pathogens. Asci are formed naked—not enclosed in a fruiting body. Yeasts traditionally have been important in the production of beer, wine, single cell protein and baker's yeast, but their role in industry has expanded to the production of citric acid, fuel alcohol, and riboflavin. Saccharomyces cerevisiae, the yeast used in baking and brewing, is an important model organism for scientists studying a wide range of genetic and physiological processes. In 1997, S. cerevisiae was the first eukaryotic organism to have its genome completely sequenced.

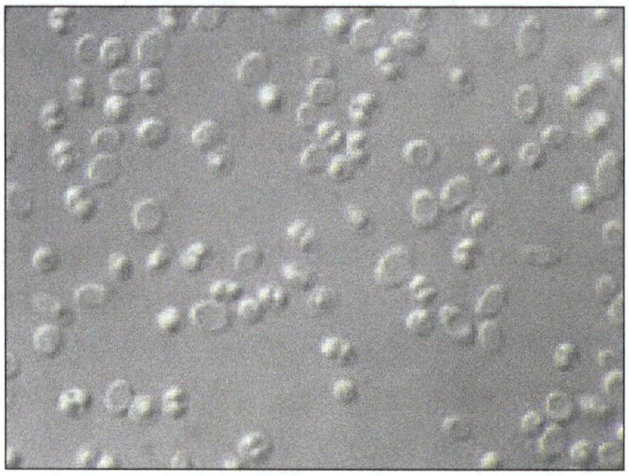

Subphylum Pezizomycotina is the largest group in the phylum, with more than 32,000 identified species that occupy a wide range of ecological niches, occurring as saprotrophs, parasites and mutualists with plants, animals and other fungi. At least 40% of the species form lichens. Three different types of asci occur in this subphylum, prototunicate, unitunicate and bitunicate.

Prototunicate asci release ascospores by breakdown of the ascus wall, whereas in the unitunicate and bitunicate asci, the ascospores are forcibly discharged. Bitunicate asci have an inner wall that balloons out from the outer wall prior to ascospore discharge and in unitunicate asci the wall layers do not separate from each other. A wide range of fruiting bodies are formed by members of subphylum Pezizomycotina, including cleistothecia, chasmothecia, apothecia, perithecia and pseudothecia. Stromata, hardened masses of hyphae on or in which perithecia or pseudothecia are formed, occur in some members of this subphylum.

Cleistothecia (sing. cleisothecium) lack a preformed opening and ascospores are released by the breakdown of the ascomatal wall. Common fungi that produce cleistothecia include the teleomorphic (sexual) states of Aspergillus and Penicillium. Species of Aspergillus are important in the production of fermented foods and beverages, including soy sauce, miso and rice wine (sake). Some species of Aspergillus infect animals, causing a disease known as aspergillosis, and others produce mycotoxins. Aflatoxin is a potent carcinogen produced by A. flavus that occurs in food made from cereals, corn and peanuts that have been colonized by the fungus. The U. S. Food and Drug Administration established a strict limit of 20 parts per billion on aflatoxin levels in food, and the U. S. spends $30-50 million annually on aflatoxin testing. Penicillium species are also used in food production. For example, the blue veins in Roquefort and Gorgonzola cheeses are due to the growth and sporulation of particular species of Penicillium, as is the white rind on the outside of Camembert cheese. The antibiotic penicillin, the "wonder drug" of the 20th century, is produced by strains of P. chrysogenum. Other species of Penicillium, such as P. digitatum and P. italicum cause the blue and green molds commonly causing rots of citrus fruits.

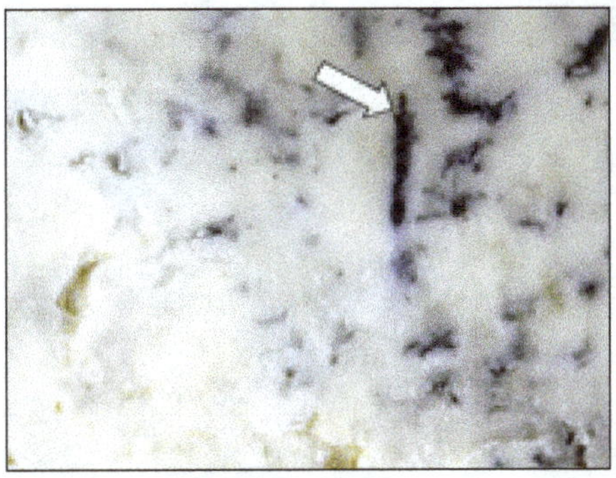

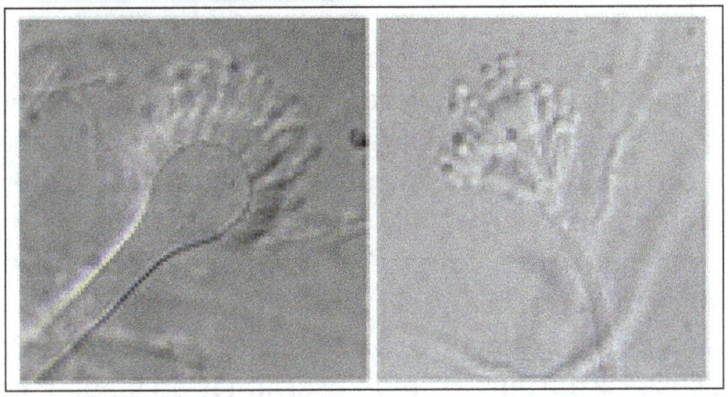

Chasmothecia (sing. chasmothecium) also lack a preformed opening, but ascospores are released by a split (or "chasm") in the ascomatal wall. The term is now used to refer to the fruiting bodies of the powdery mildew fungi in order Erysiphales.

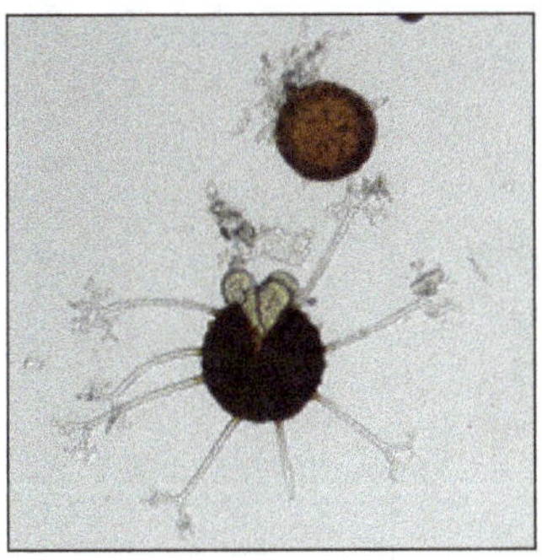

Apothecia (sing. apothecium) are exposed, often cup-shaped fruiting bodies, but can take on a variety of forms, including those found in the morels. Apothecia-forming fungi are also called "cup fungi" or discomycetes. Some important groups of plant pathogens that form apothecia include species of Monilinia, and Sclerotinia.

Perithecia (sing. perithecium) are enclosed ascomata with a preformed opening (ostiole) through which ascospores are discharged. Most fungi producing perithecia also have unitunicate asci and are classified in Sordariomycetes, one of the largest classes of Ascomycota with more than 3,000 described species. These fungi have also been called pyrenomycetes. Members of this group are common in nearly all ecosystems, where they occur as saprotrophs, endophytes of plants, or pathogens of plants, animals and other fungi. A large number of economically important plant pathogens belong to Sordariomycetes, including those that cause anthracnose diseases (Glomerella cingulata), blasts (Magnaporthe oryzae, rice blast pathogen), blights (Cryphonectria parasitica, chestnut blight), ergot (Claviceps purpurea), and Fusarium head blight (scab) of small grains (Gibberella zeae).

Pseudothecia (sing. pseudothecium) look similar to perithecia, but they differ in development. Asci form in locules (openings) inside vegetative fungal tissue called ascostroma; this group has been called loculoascomycetes, but is now placed in class Dothideomycetes. Other characteristics of Dothideomycetes include the formation of bitunicate asci, and many members of this group produce darkly pigmented, multiseptate asospores or conidia. Similar to the Sordariomycetes, members of Dothideomycetes occur in a wide range of habitats as saprotrophs and associate with plants as pathogens, endophytes and growing on the surface of plants as epiphytes. Examples of well-known plant pathogens belonging to this group include Venturia inaequalis (apple scab) and Mycosphaerella fijiensis (Black Sigatoka disease of banana); some of the common asexually reproducing fungi that belong in Dothideomycetes are Alternaria, Cladosporium, *Phoma*, and Stemphylium.

Most of the lichen-forming members of Ascomycota belong in class Lecanoromycetes. This is the largest class of fungi, with over 13,500 described species. Most of the members of this class produce apothecial fruiting bodies. The majority of lichenized fungi form a symbiotic association with green algae, and approximately 10% are associated with cyanobacteria. The lichen thallus produces a wide range of secondary metabolites that are of biological and ecological importance. The lichen thallus is able to grow under a range of adverse conditions and some can survive for hundreds of years. Lichens are found in a wide range of habitats from the Arctic to Antarctic, including some species that can grow in aquatic and marine environments.

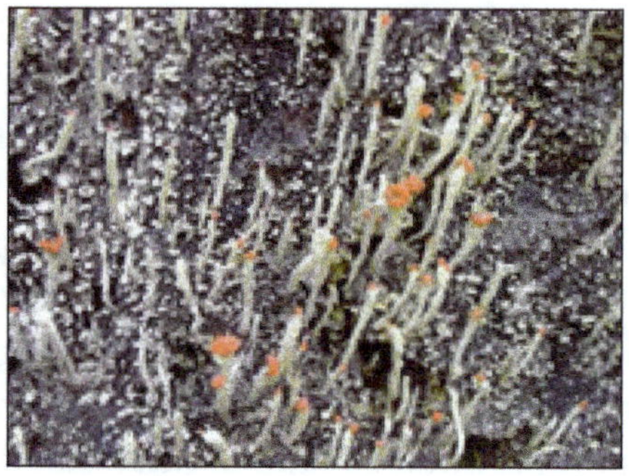

Phylum Basidiomycota

Phylum Basidiomycota represents the second largest phylum of fungi, with nearly 30,000 described species. Members of phylum Basidiomycota produce basidiospores on a typically club-shaped structure called a basidium. Characteristic of the mycelium of many members of Basidiomycota is the presence of clamp connections and dolipore septa.

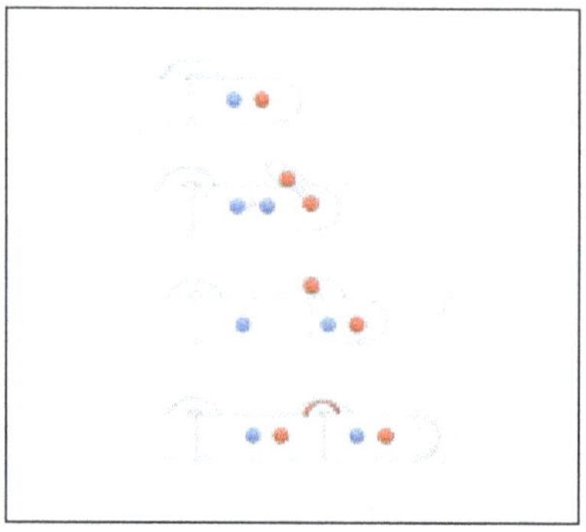

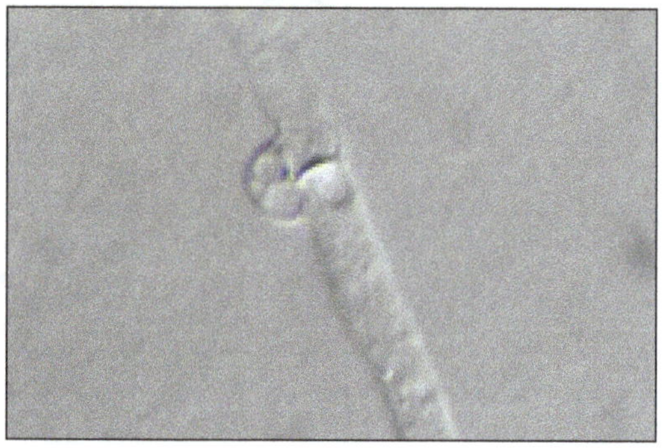

Three main lineages are recognized in phylum Basidiomycota: subphyla Ustilaginomycotina, Pucciniomycotina, and Agaricomycotina. Ustilaginomycotina and Pucciniomycotina are composed mostly of plant parasitic species; known as smut and rust fungi, respectively, characterized by a state that produces thick-walled teliospores. The most extensively studied members of Ustilaginomycotina are species of Tilletia and Ustilago. Ustilago maydis, which causes corn smut, produces conspicuous tumor-like growths on infected host plants. These structures eventually become filled with dark teliospores and are considered a delicacy in Mexico called "cuitlacoche." The most economically important species of Tilletia are the wheat bunts (T. caries and T. laevis—common bunt; T. contraversa—dwarf bunt; and T. indica—Karnal bunt), plant pathogens that convert host ovaries into masses of dark, thick-walled teliospores that smell like rotting fish.

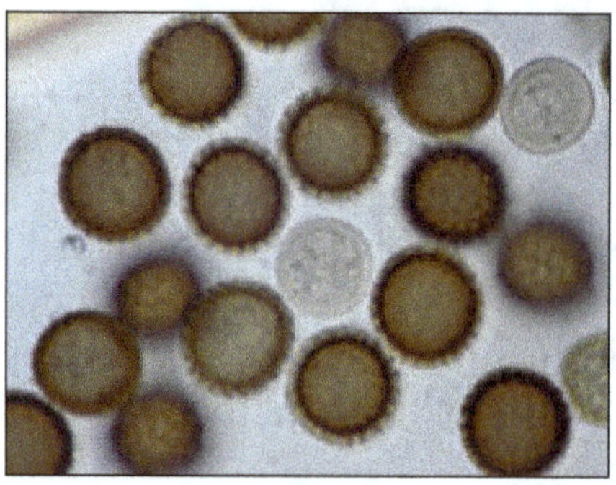

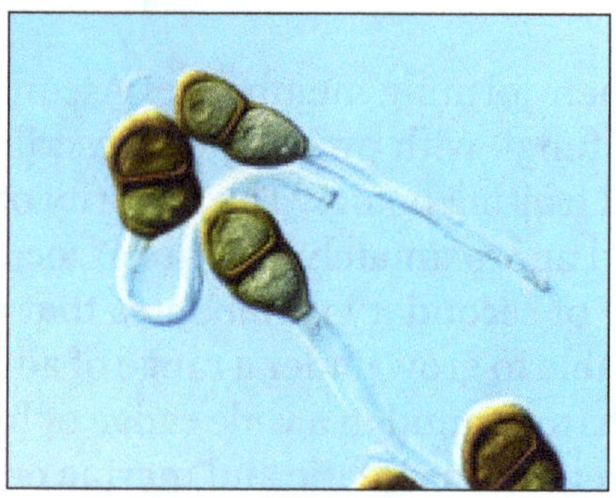

Subphylum Pucciniomycotina include the group of plant parasites called rust fungi. The rust fungi are remarkable in having as many as five distinct types of spores in a single life cycle (spermatia, aeciospores, urediniospores, teliospores, and basidiospores). Rust fungi that produce all five spore states are macrocyclic, those that do not form uredinospores are demicyclic, and those that do not form urediniospores and aeciospores are microcyclic. Rust fungi may complete the life cycle on one host (autoecious rusts) or require two unrelated alternate hosts for completion of the life cycle (heteroecious rusts). The most widely cited example of a macrocyclic, heteroecious rust is Puccinia graminis (black stem rust), which forms two spore states (uredinia and telia) on cultivated wheat and two different spore states (spermagonia and aecia) on barberry leaves. The fifth state, the probasidium producing basidiospores, is formed upon germination of the teliospores.

Subphylum Agaricomycotina, previously known as the Hymenomycetes, includes the morphologically diverse group of fungi that produce basidia in various types of fruiting bodies (general life cycle). This group includes the fungi commonly known as mushrooms, puffballs, shelf fungi, stinkhorns, jelly fungi and bird's nest fungi. Many species are saprotrophic, utilizing dead plant material including woody substrates. Some of these saprotrophic species are cultivated for food, for example, the common button mushroom (Agaricus bisporus), oyster mushrooms (Pleurotus ostreatus), and shiitake (Lentinula edodes). Other members of this group are important ectomycorrhizal fungi, forming mutualistic associations with the roots of a wide range of trees. Some fruiting bodies produced by ectomycorrhizae are considered choice edibles, for example, chanterelles (Cantharellus cibarius and other species), porcini (Boletus edulis), and the American matsutake (Tricholoma magnivelare). A few members of this group are economically important plant parasites, e. g., species of Armillaria and Rhizoctonia.

Phylum Glomeromycota

The arbuscular mycorrhizal (AM) fungi, long considered to belong in the Zygomycota, are now recognized as comprising a distinct phylum, Glomeromycota. This is an ancient group of fungi, recognizable in the fossil record dating back at least 400 million years. The AM fungi form obligate, mutualistic associations, called endomycorrhizae, with the roots of most (~ 80%) vascular plants. Only a small number (~ 160) of species is recognized in the phylum. One of the most distinctive features of these fungi is the highly branched arbuscules formed inside the cortical cells of host roots; arbuscules are the point of exchange between fungus and plant, where carbohydrates produced by the plant are acquired by the fungus, and where nitrogen, phosphorous, and other minerals acquired by the mycelium of the fungus are transferred to the plant. The fungus acquires as much as 20-40% of the photosynthate produced by the plant. Some AM fungi also produce storage structures inside plant roots called vesicles. Endomycorrhizal fungi produce an extensive network of hyphae outside the roots (extra radical hyphae). The extra radical hyphae act like an extension of the plant roots, increasing the plant's access to water and soil minerals, particularly phosphorous and nitrogen. The fungus is also able to access phosphate not otherwise available to plants, for example from organic matter by production of acid phosphatases. Reproduction of AM fungi is by thick-walled spores ranging in size from 40–800 µm in diameter, each of which can contain hundreds or thousands of nuclei. Spores may be formed singly or in clusters, and the mycelium of AM fungi is coenocytic. Sexual reproduction is not known to occur in this phylum.

Phylum Chytridiomycota

"Chytrids" are a small group of fungi with approximately 900 identified species occurring in a wide range of aquatic and terrestrial habitats around the world. The feature that is shared by all members of this phylum is the formation of zoospores with one posteriorly directed, whiplash flagellum. A few Chytrids are economically important plant pathogens, e. g., Synchytrium endobioticum, which causes the black wart disease of potato, others are vectors of plant viruses (Olpidium), but most are saprotrophs using substrates such as cellulose, chitin, and keratin as a food source. The frog chytrid, Batrachochytrium dendrobatidis, has been implicated as a major factor in population declines of frogs and other amphibians around the world.

Phylum Zygomycota

This phylum contains approximately 900 identified species divided amongst two ecologically distinct classes, Zygomycetes and Trichomycetes. Hibbett et al indicate the phylum is polyphyletic and further work is needed to clarify relationships of fungi traditionally considered in Zygomycota. The most commonly encountered Zygomycetes are members of orders Mortierellales and Mucorales. Many members of these two orders are saprotrophs with rapidly growing, coenocytic mycelium. The sexual reproductive state is the zygospore, but many of these fungi produce a large number of readily dispersed asexual spores called sporangiospores. Members of order Mucorales, commonly called mucoraceous fungi, are common in soil, dung, plant material, and other types of organic matter. Some mucoraceous fungi are plant or animal pathogens, and others are used in the production of Asian foods such as tempeh. Species of Mucor and Rhizopus are commonly isolated from decaying organic matter and can cause decay diseases of fleshy fruits, vegetables, and sunflower peduncles.

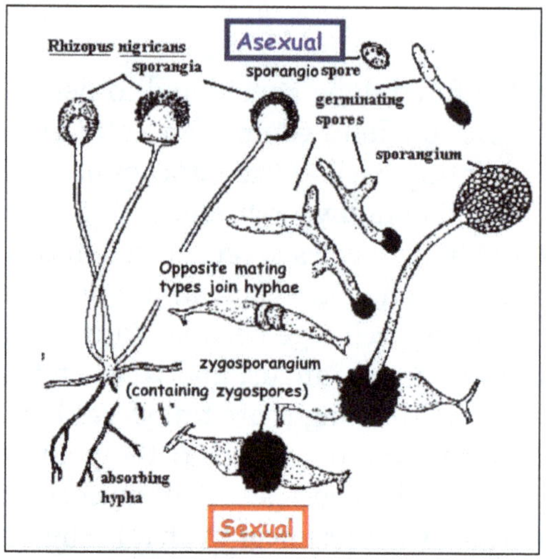

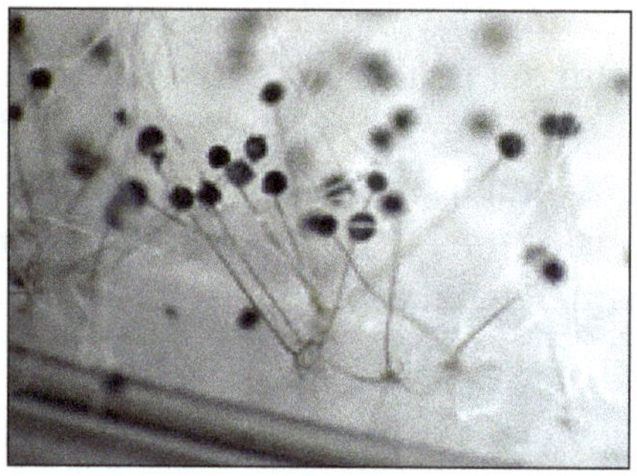

Species of Pilobolus are among the first fungi observed growing on herbivore dung incubated in moist chambers. Other Zygomycetes are associated with animals. For example, some species of Rhizopus and Mucor cause zygomycosis in immunocompromised humans. Entomophthorales, as the name suggests, include parasites of insects and other animals. Members of class Trichomy-cetes live in the guts of insects, millipedes, and crustaceans, but cause little or no harm to their hosts.

 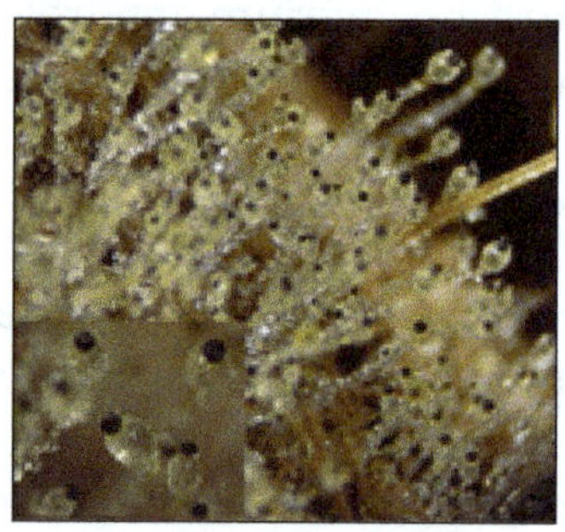

Fungal-like Organisms Studied by Plant Pathologists and Mycologists.

Oomycetes

Oomycetes are fungal-like organisms that form zoospores with two flagella—a whiplash flagellum that is directed backwards and propels the zoospore, and a tinsel flagellum adorned with hairs that is directed forward, pulling the zoospore. The cell walls of Oomycetes contain cellulose, rather than chitin, and glucans. Another characteristic of Oomycetes is the formation of an oospore, a thick-walled, resistant propagule which is the result of sexual reproduction. Oomycetes belong in

Kingdom Straminipila, also known as Chromista. In addition to Oomycetes, this kingdom includes diatoms, golden and brown algae, a type of algae called cryptomonads, and two other groups of organisms studied by mycologists in phyla Labyrinthulomycota and Hyphochytriomycota. The tinsel type flagellum is a characteristic of all members of Kingdom Straminipila; hence it is also called a straminipilous flagellum.

Phytophthora ramorum causes sudden oak death and ramorum blight. Pythium species cause damping diseases under extremely wet conditions. Seedlings are particularly vulnerable to attack by damping-off pathogens because their tissues are soft and easily invaded. Seedlings may be killed before or after they emerge from the soil. Downy mildews (Peronosporales) are biotrophic Oomycetes that are characterized by the formation of white, downy sporangiophores on the surface of infected hosts. The white rusts (Albugo spp.) produce chains of sporangia that erupt through the cuticle of infected hosts. Albugo species parasitize crucifers and produce blister-like pustules filled with sporangia, which will germinate to produce motile zoospores. White rusts can also cause infected stems to grow in a contorted or twisted manner.

Members of Saprolegniales are called water molds. Many of these produce fast-growing, robust hyphae on organic matter in aquatic environments, but some species of Saprolegnia are parasitic on fish and fish eggs.

Other Groups

Labyrinthulids (phylum Labyrinthulomycota), a small group of Straminipila, include organisms that cause rapid blight of turf grass and the wasting disease of eelgrass. Labyrinthulids have a unique manner of movement—their microscopic, football-shaped cells produce an ectoplasmic net through which the cells glide. The slow, gliding movement of the cells within the ectoplasmic net can be observed under the microscope.

Hyphochytrids (phylum Hyphochytridomycota, Kingdom Straminipila) are similar to chytrids in appearance, as their name suggests, and produce zoospores with a single anterior tinsel flagellum. Hyphochytrids are one of the smallest groups of fungal-like organisms, both in size and in number of species with only 23 known. Some hyphochytrids are known to parasitize algae, spores of AM fungi, and oospores of Oomycetes.

Slime Molds

Slime molds are organisms that have a trophic (feeding) stage in their life cycle that lacks a cell wall, either uninucleate (amoeba) or multinucleate (plasmodium). The lack of a cell wall facilitates engulfment of food, in contrast to true fungi that must absorb their nutrients through a cell wall. The slime molds are now included in the Amoebozoa.

Four groups of slime molds are recognized—plasmodial slime molds (Myxomycota), cellular slime molds (Dictyosteliomycota and Acrasiomycota) and endoparasitic slime molds (Plasmodiophoromycota). We will briefly cover plasmodial slime molds and endoparasitic slime molds. The plasmodial slime molds are most commonly found in temperate forests, where they occur on plant litter, tree bark, and other types of plant material. They produce a multinucleate trophic stage lacking a cell wall called a plasmodium that moves over and through decaying organic matter, engulfing bacteria, fungi, and other microorganisms. The most conspicuous stage of the plasmodial slime mold is the fruiting structures, called sporophores, which are often brightly colored and visible to the naked eye. One of the most common slime molds in temperate regions is Fuligo septica. The sporophores of this slime mold are often found in ornamental bark and mulch, and look more like an animal's vomit than the fruiting structure of a living organism, thus earning the nickname: "dog vomit slime". The presence of slime molds in landscaping occasionally prompts calls to plant disease clinics, however, none of the plasmodial slime molds are known to be plant or animal parasites, and are of no known economic importance except as model organisms for research.

Members of phylum Plasmodiophoromycota are biotrophic parasites that produce their plasmodial stage inside the cells of plants, algae, diatoms, and Oomycetes. Several members of this phylum are economically important plant parasites, including Plasmodiophora brassicae, which causes clubroot of crucifers and Spongospora subterranea, which causes powdery scab of potato. Polymyxa graminis is a vector for soilborne wheat mosaic virus, an economically important disease of wheat. Members of this phylum produce cysts inside host cells; the cysts are released when the plant tissue breaks down, and germinate to release a zoospore that infects the host by injecting its cytoplasm into a host cell. Infected host tissue may become greatly swollen as in clubroot of crucifers.

Protozoa

Organisms known as protozoa include a wide range of organisms, most of which are free-living single-celled eukaryotes. Therefore, protozoa fit into the Domain Eukarya. Although the different phyla of the kingdom Protista are not closely related, they are nonetheless classified together because of their large differences from the other kingdoms of plants, animals and fungi. The name "protozoa" has a dynamic history, at one time including only the "animal-like" unicellular forms of life. Today, these heterotrophic protozoa are lumped together with the autotrophic algae and other simple forms of life into the Kingdom Protista. While all protozoa are eukaryotes, not all reproduce with the standard model of mitosis that is seen in higher animal cells. Many have complex cellular division that resembles binary fission in bacteria, on a larger scale. Some phyla in the Kingdom Protista are autotrophic cells, containing chloroplasts which can produce sugars from sunlight.

Examples of Protozoa

Malaria

Malaria is a disease that affects hundreds of millions of people worldwide, every year. It is estimated that malaria kills three million people a year, half of which are young children. Malaria is caused by several protozoa in the genus Plasmodium. This particular protozoa carries out life cycles in both the digestive systems of mosquitos, and in human blood cells. The unicellular protozoans enter the bloodstream of humans during a mosquito bite. They find an uninfected red blood cell, and use a special combination of organelles at the end of their cell to force the cell to draw the parasite in. This special group of organelles, known as the apical complex, allows the parasitic protozoa to hide within blood cells. This makes Plasmodium and other parasitic Apicomplexans very hard to kill, as the blood cells would have to be destroyed as well.

Red Tide

An interesting phenomenon in the ocean is also caused by another phylum of protozoa. The Dinoflagellates are a group of protozoa that exist mainly as free-swimming unicellular organisms, while other exists symbiotically within certain invertebrates. Groups of dinoflagellates that are free swimming can undergo drastic periods of reproduction, known as blooms. If a bloom produces enough dinoflagellates, the ocean water will become pink or even red because of their density. Unfortunately for any organism that passes through the bloom, dinoflagellates produce a toxin. When congregated in enormous densities, this neurotoxin can paralyze and kill almost any organism in the bloom. Fish, crustaceans, and mollusks that eat the bloom may store the toxins in their tissues and become poisonous to people who eat them. Paralytic Shellfish Poisoning is the caused by dinoflagellate toxins in high concentration. Simple protozoa, in high numbers, can cause muscle paralysis and respiratory failure in humans, possibly leading to death.

Types of Protozoa

There are many phyla in the Domain Protista. While this represents a wide variety of them, it is not close to all of them. Part of the reason there is so much diversity in this domain is that the phyla are not closely related.

Phylum Euglenida

The Euglenida are a phylum of protozoa recognized by the pellicle that gives them shape and the flagella which they use for locomotion. The pellicle is a shell of sorts that exists beneath the cell membrane. It is composed of strips of proteins, which interlock for support. Some organisms in the Euglenida are photosynthetic, and contain chloroplasts. Others obtain food from dissolved nutrients in the environment, while still others are parasitic.

Phylum Kinetoplastida

Closely related to the Euglenida, the Kinetoplastida are also protected by a pellicle, although it is made exclusively of microtubules. Organisms in the Kinetoplastida share the unique characteristic of having a single, much enlarged and elongated, mitochondrion. Typically, cells have many small mitochondria, as opposed to one large one. The Kinetoplastida includes many parasitic organisms that cause disease in humans. Of these, leishmaniosis is the most notable, affecting over a million people a year. However, advances in medical treatments saves most of those infected, and only around 1000 people die annually from these protozoa.

Phylum Ciliophora

Instead of flagella for locomotion, organisms in the phylum Ciliophora use much smaller structures called cilia. The cilia of these organisms cover their entire cell, and work together to propel the cell forward. Much like the individual paddles of a row boat, each cilia gives a forward moving power stoke, then whips back to the starting position in the recovery stroke. Organisms in the Ciliophora include a wide variety of body plans, including free-swimming organisms and sessile organisms that use their cilia to filter food from the water. Most ciliates exist on the bottom of marine environments, known as the benthic zone. However, these protozoa have also specialized as parasites in the digestive tracts of larger organisms.

Phylum Apicomplexa

The Apicomplexa include mostly parasitic organisms that exist solely within their host organisms. These organisms all have apical complexes, which allow them to gain entry into their host's cells. Inside the cells, the parasites are much more resistant and have better access to nutrients. Many of these parasites can also change the proteins exposed on the surface of their cells. This helps them "hide" from the immune system cells, which recognize foreign cells on the surface of invading pathogens. This makes it very difficult to treat medically.

Phylum Dinoflagellata

The dinoflagellates use flagella to locomote and are also covered by a pellicle. Unlike the two previous groups with pellicles, the dinoflagellates use a series of vesicles beneath the cell membrane to create a rigid form. Some dinoflagellates even fill their vesicles with polysaccharides, like cellulose, to create a form of armor that protects their cells.

Phylum Stramenopila

The Stramenopila includes a wide variety of organisms, from the shelled diatoms to brown and golden algae. Protozoa in this phylum have shells, scales, or tests that support the cell. These protective covering vary widely across the phylum. The diatoms tend to make ornate and complex tests of silicate, which is also used to make glass. Others form their shells from calcium carbonate or protein.

Phylum Rhizopoda

The phylum Rhizopoda contains the amoebas. The small, unicellular protozoa are some of the only protozoa that do not have any sort of hard covering. The amoebas move by extending their cytoplasm into the environment. These extensions are called pseudopodia.

Phylum Actinopoda

The organisms in Actinopoda have characteristic axopodia, which are sharp spines that extend from the cell and are covered in cytoplasm. They are used in locomotion and feeding. The axopodia also produce beautiful shapes, as seen in the organism below:

Figure: Radiolarian.

Phylum Granuloreticulosa

The Granuloreticulosa includes a variety of organism that has immense industrial value. The tests produced by these protozoa aggregate in huge numbers on the bottom of marine environments.

Over time they fossilize together and become substances like chalk, limestone, and marble. Even the pyramids of Egypt were built from stones that originated from the shells of these protozoans.

Phylum Diplomonodida

The Diplomonodida are a phylum of organisms that have many flagella for locomotion. The average number is around 8. While there are many species that vary in their place in the ecosystem, the most well-known of these protozoa is the genus Giardia, which can be consumed in unclean water and cause flu-like symptoms and diarrhea in humans.

Phylum Parabasilida

These organisms contain many flagella, up to thousands on a single cell, and have a special fiber that attached the Golgi apparatus to the base of the flagella. Many parabasilids exists as symbiotic protozoa in the digestive tracts of insects, particularly those that eat wood. These protozoa are able to secrete an enzyme that can break down cellulose.

Other Protozoa

Several other phyla of protozoa exist. Like all other protozoans, these organisms lack specialized tissues or layers of cells. Most are unicellular, although some exist in colonies or fibers. Due to their small size adaptability, the protozoa have occupied nearly every environment on the planet.

Viruses

Viruses are the smallest of all the microbes. They are said to be so small that 500 million rhinoviruses (which cause the common cold) could fit on to the head of a pin. They are unique because they are only alive and able to multiply inside the cells of other living things. The cell they multiply in is called the host cell.

A virus is made up of a core of genetic material, either DNA or RNA, surrounded by a protective coat called a capsid which is made up of protein. Sometimes the capsid is surrounded by an additional spikey coat called the envelope. Viruses are capable of latching onto host cells and getting inside them.

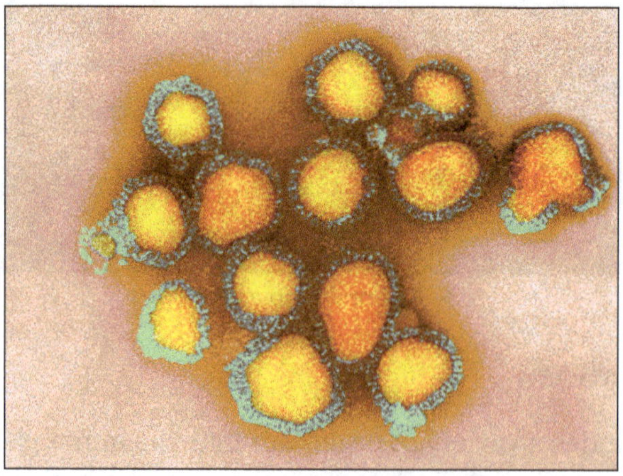

Figure: H3N2 influenza virus particles, coloured transmission electron micrograph (TEM).

Each virus consists of a nucleocapsid (protein coat) that surrounds a core of RNA (ribonucleic acid) genetic material. Surrounding the nucleocapsid is a lipid envelope that contains the glycoprotein spikes haemagglutinin (H) and neuraminidase (N). These viruses were part of the Hong Kong Flu pandemic of 1968-1969 that killed approximately one million people worldwide. H3N2 viruses are able to infect birds and mammals as well as humans. They often cause more severe infections in the young and elderly than other flu strains and can lead to increases in hospitalizations and deaths.

Viruses only exist to make more viruses. The virus particle attaches to the host cell before penetrating it. The virus then uses the host cell's machinery to replicate its own genetic material. Once replication has been completed the virus particles leave the host by either budding or bursting out of the cell (lysis).

Budding

As the newly formed viral particle pushes against the host cell's plasma membrane a portion adheres to it. The plasma membrane envelops the virus and becomes the viral envelope. The virus is released from the cell. This process slowly uses up the host's cell membrane and usually leads to cell death.

Lysis

The virus particles burst out of the host cell into the extracellular space resulting in the death of the host cell. Once the virus has escaped from the host cell it is ready to enter a new cell and multiply.

Classification of Virus

Viruses are classified on the basis of morphology, chemical composition, and mode of replication. The viruses that infect humans are currently grouped into 21 families, reflecting only a small part of the spectrum of the multitude of different viruses whose host ranges extend from vertebrates to protozoa and from plants and fungi to bacteria.

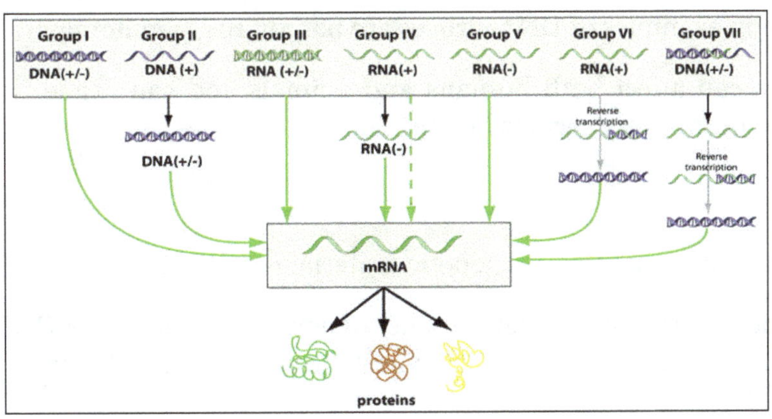

The following properties have been used as a basis for the classification of viruses:

- Virion morphology, including size, shape, type of symmetry, presence or absence of peplomers, and presence or absence of membranes.

- Virus genome properties, including type of nucleic acid (DNA or RNA), size of genome in kilobases (kb) or kilobase pairs (kbp), strandedness (single or double), whether linear or circular, sense (positive, negative, ambisense), segments (number, size), nucleotide sequence, G + C content, and presence of special features (repetitive elements, isomerization, 5′-terminal cap, 5′-terminal covalently linked protein, 3′-terminal poly(A) tract).
- Genome organization and replication, including gene order, number and position of open reading frames, a strategy of replication (patterns of transcription, translation), and cellular sites (accumulation of proteins, virion assembly, virion release).
- Virus protein properties, including number, size, and functional activities of structural and nonstructural proteins, amino acid sequence, modifications (glycosylation, phosphorylation, myristylation), and special functional activities (transcriptase, reverse transcriptase, neuraminidase, fusion activities).
- Antigenic properties.
- Physicochemical properties of the virion, including molecular mass, buoyant density, pH stability, thermal stability, and susceptibility to physical and chemical agents, especially ether and detergents.
- Biologic properties, including natural host range, mode of transmission, vector relationships, pathogenicity, tissue tropisms, and pathology.

On the basis of genetic material present viruses are small, nonliving parasites, which cannot replicate outside of a host cell. A virus consists of genetic information — either DNA or RNA — coated by a protein. Accordingly, they are classified as DNA viruses and RNA viruses. The nucleic acid may be single or double stranded, circular or linear, segmented or unsegmented.

DNA Viruses

- As their name implies, DNA viruses use DNA as their genetic material.
- Some common examples of DNA viruses are parvovirus, papillomavirus, and herpes virus.
- DNA viruses can affect both humans and animals and can range from causing benign symptoms to posing very serious health.

RNA Viruses

- The virus that possesses RNA as genetic material is called RNA virus.
- Rotavirus, polio virus, yellow fever virus, dengue virus, hepatitis C virus, measles virus, rabies virus, influenza virus and Ebola virus are examples of RNA virus.

DNA-RNA Viruses

- The RNA tumor viruses called Leuko viruses and Rous's viruses unusually contain both DNA and RNA as genetic material.

On the basis of the presence of a number of strands:

- Double-stranded DNA: It is found in pox viruses, the bacteriophages T2, T4, T6, T3, T7 and Lamda, herpes viruses, adenoviruses etc.
- Single-stranded DNA: It is found in bacteriophages φ, X, 74 bacteriophages.
- Double-stranded RNA: It has been found within viral capsid in the reoviruses of animals and in the wound tumor virus and rice dwarf viruses of plants.
- Single-stranded RNA: It is found in most of the RNA viruses e.g. tobacco mosaic virus, influenza virus, and poliomyelitis, and bacteriophage MS-2, avian leukemia virus.

On the Basis of Presence of Envelope

The envelope is a lipid-containing membrane that surrounds some virus particles. It is acquired during viral maturation by a budding process through a cellular membrane. Virus encoded glycoproteins are exposed on the surface of the envelope. These projections are called peplomers.

Enveloped Virus

- DNA viruses: Herpesviruses, Poxviruses, Hepadnaviruses.
- RNA viruses: Flavivirus, Toga virus, Coronavirus, Hepatitis D, Orthomyxovirus, Paramyxovirus, Rhabdovirus, Bunyavirus, Filovirus.

Retroviruses

Non-Enveloped Virus

- DNA viruses- parvovirus, adenovirus and papovavirus.
- RNA viruses- Picornavirus, Hepatitis A virus and Hepatitis E virus.

Virus Classification by Capsid Structure

- Naked icosahedral: Hepatitis A virus, polioviruses.
- Enveloped icosahedral: Epstein-Barr virus, herpes simplex virus, rubella virus, yellow fever virus, HIV-1.
- Enveloped helical: Influenza viruses, mumps virus, measles virus, and rabies virus.
- Naked helical: Tobacco mosaic virus.
- Complex with many proteins: Some have combinations of icosahedral and helical capsid structures. Herpesviruses, smallpox virus, hepatitis B virus, T4 bacteriophage.

On the Basis of Shapes of the Viruses

- Most of the animal viruses are roughly spherical with some exceptions.

- Rabies virus: Bullet shaped.
- Ebola virus: Filamentous shaped.
- Poxvirus: Brick shaped.
- Adenovirus: Space vehicle shaped.

Classification of Virus on the Basis of Structure

- Cubical virus: They are also known as icosahedral symmetry virus. E.g. Reo virus, Picorna virus.
- Spiral virus: They are also known as helical symmetry virus. E.g. Paramyxovirus, orthomyxovirus.
- Radial symmetry virus: E.g. Bacteriophage.
- Complex virus: E.g. Pox virus.

On the Basis of the Type of Host

The virus can be classified on the basis of the type of host. They are:

- Animal viruses.
- Plant viruses.
- Bacteriophage.

Animal Viruses

The viruses which infect and live inside the animal cell including man are called animal viruses. E.g. influenza virus, rabies virus, mumps virus, poliovirus etc. Their genetic material is RNA or DNA.

Plant Viruses

The viruses that infect plants are called plant viruses. Their genetic material is RNA which remains enclosed in the protein coat. Some plant viruses are tobacco mosaic virus, potato virus, beet yellow virus and turnip yellow virus etc.

Bacteriophages

Viruses which infect bacterial cells are known as bacteriophage or bacteria eaters. They contain DNA as genetic material. There are many varieties of bacteriophages. Usually, each kind of bacteriophage will attack only one species or only one strain of bacteria.

Classification of Virus on the Basis of Mode of Transmission

- Virus transmitted through respiratory route: E.g., Swine flu, Rhino virus.

- Virus transmitted through faeco-oral route: E.g. Hepatitis A virus, Polio virus, Rota virus.
- Virus transmitted through sexual contacts: E.g. Retro virus.
- Virus transmitted through blood transfusion: E.g. Hepatitis B virus, HIV.
- Zoonotic virus: Virus transmitted through biting of infected animals; Eg. Rabies virus, Alpha virus, Flavi virus.

Classification of Virus on the Basis of Replication Properties and Site of Replication

- Replication and assembly in cytoplasm of host: All RNA virus replicates and assembles in cytoplasm of host cell except Influenza virus.
- Replication in nucleus and assembly in cytoplasm of host: Influenza virus, Pox virus.
- Replication and assembly in nucleus of host: All DNA viruses replicate and assemble in nucleus of host cell except Pox virus.
- Virus replication through ds DNA intermediate: All DNA virus, Retro virus and some tumor causing RNA virus replicates through ds DNA as intermediates.
- Virus replication through ss RNA intermediate: All RNA virus except Reo virus and tumor causing RNA viruses.

Baltimore Classification

The most commonly used system of virus classification was developed by Nobel Prize-winning biologist David Baltimore in the early 1970s. In addition to the differences in morphology and genetics mentioned above, the Baltimore classification scheme groups viruses according to how the mRNA is produced during the replicative cycle of the virus. These groups are as follows:

- Group I: Viruses contain double-stranded DNA (dsDNA) as their genome. Their mRNA is produced by transcription in much the same way as with cellular DNA.
- Group II: Viruses have single-stranded DNA (ssDNA) as their genome. They convert their single-stranded genomes into a dsDNA intermediate before transcription to mRNA can occur.
- Group III: Viruses use dsRNA as their genome. The strands separate, and one of them is used as a template for the generation of mRNA using the RNA-dependent RNA polymerase encoded by the virus.
- Group IV: Viruses have ssRNA as their genome with a positive polarity. Positive polarity means that the genomic RNA can serve directly as mRNA. Intermediates of dsRNA, called replicative intermediates, are made in the process of copying the genomic RNA. Multiple, full-length RNA strands of negative polarity (complementary to the positive-stranded genomic RNA) are formed from these intermediates, which may then serve as templates for the production of RNA with positive polarity, including both full-length genomic RNA and shorter viral mRNAs.

- Group V: Viruses contain ssRNA genomes with a negative polarity, meaning that their sequence is complementary to the mRNA. As with Group IV viruses, dsRNA intermediates are used to make copies of the genome and produce mRNA. In this case, the negative-stranded genome can be converted directly to mRNA. Additionally, full-length positive RNA strands are made to serve as templates for the production of the negative-stranded genome.
- Group VI: Viruses have diploid (two copies) ssRNA genomes that must be converted, using the enzyme reverse transcriptase, to dsDNA; the dsDNA is then transported to the nucleus of the host cell and inserted into the host genome. Then, mRNA can be produced by transcription of the viral DNA that was integrated into the host genome.
- Group VII: Viruses have partial dsDNA genomes and make ssRNA intermediates that act as mRNA, but are also converted back into dsDNA genomes by reverse transcriptase, necessary for genome replication.

Prions

A prion is an infectious agent composed of protein in a misfolded form. This is the central idea of the Prion Hypothesis, which remains debated. This is in contrast to all other known infectious agents (virus /bacteria/fungus/parasite) who must contain nucleic acids (either DNA, RNA, or both). The word prion, coined in 1982 by Stanley B. Prusiner, is derived from the words protein and infection. Prions are responsible for the transmissible spongiform encephalopathies in a variety of mammals, including bovine spongiform encephalopathy (BSE, also known as "mad cow disease") in cattle and Creutzfeldt–Jakob disease (CJD) in humans. All known prion diseases affect the structure of the brain or other neural tissue, are currently untreatable and universally fatal.

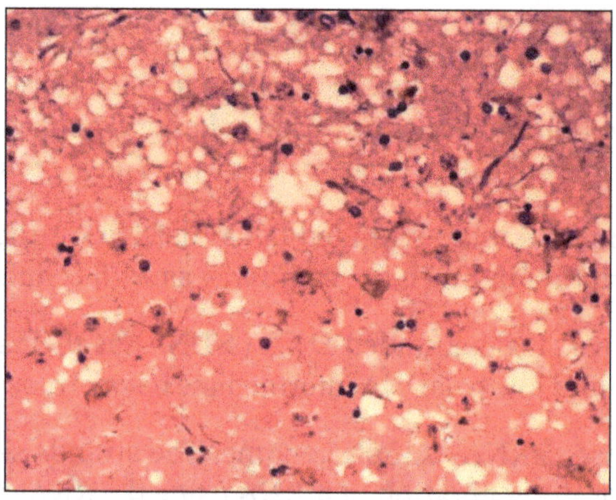

Figure: Prion-affected tissue.

This micrograph of brain tissue reveals the cytoarchitectural histopathologic changes found in bovine spongiform encephalopathy. The presence of vacuoles, i.e. microscopic "holes" in the gray matter, gives the brain of BSE-affected cows a sponge-like appearance when tissue sections are examined in the lab.

Prions propagate by transmitting a misfolded protein state. When a prion enters a healthy organism, it induces existing, properly folded proteins to convert into the disease-associated prion form; it acts as a template to guide the misfolding of more proteins into prion form. These newly formed prions can then go on to convert more proteins themselves triggering a chain reaction. All known prions induce the formation of an amyloid fold, in which the protein polymerises into an aggregate consisting of tightly-packed beta sheets. Amyloid aggregates are fibrils, growing at their ends, and replicating when breakage causes two growing ends to become four growing ends.

The incubation period of prion diseases is determined by the exponential growth rate associated with prion replication, which is a balance between the linear growth and the breakage of aggregates. Propagation of the prion depends on the presence of normally-folded protein in which the prion can induce misfolding; animals which do not express the normal form of the prion protein cannot develop nor transmit the disease.

All known mammalian prion diseases are caused by the so-called prion protein, PrP. The endogenous, properly-folded form is denoted PrPC (for Common or Cellular) while the disease-linked, misfolded form is denoted PrPSc (for Scrapie, after one of the diseases first linked to prions and neurodegeneration.). The precise structure of the prion is not known, though they can be formed by combining PrPC, polyadenylic acid, and lipids in a Protein Misfolding Cyclic Amplification (PMCA) reaction. Proteins showing prion-type behavior are also found in some fungi, which has been useful in helping to understand mammalian prions. Fungal prions do not appear to cause disease in their hosts.

The first hypothesis that tried to explain how prions replicate in a protein-only manner was the heterodimer model. This model assumed that a single PrPSc molecule binds to a single PrPC molecule and catalyzes its conversion into PrPSc. The two PrPSc molecules then come apart and can go on to convert more PrPC. However, a model of prion replication must explain both how prions propagate, and why their spontaneous appearance is so rare. Manfred Eigen showed that the heterodimer model requires PrPSc to be an extraordinarily effective catalyst, increasing the rate of the conversion reaction by a factor of around 1015. This problem does not arise if PrPSc exists only in aggregated forms such as amyloid, where cooperativity may act as a barrier to spontaneous conversion. What is more, despite considerable effort, infectious monomeric PrPSc has never been isolated.

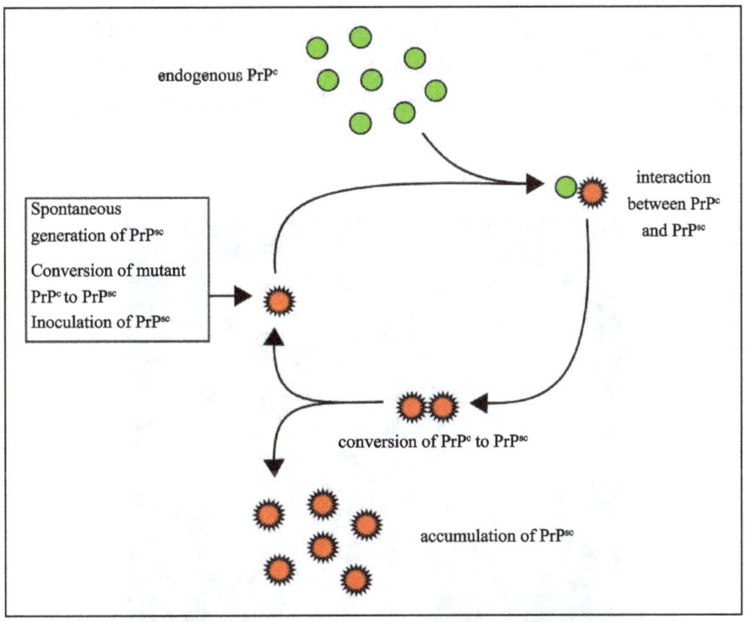

Figure: Diagram demonstrating how prion numbers increase: Heterodimer model of prion propagation.

An alternative model assumes that PrPSc exists only as fibrils, and that fibril ends bind PrPC and convert it into PrPSc. If this were all, then the quantity of prions would increase linearly, forming ever longer fibrils. But exponential growth of both PrPSc and of the quantity of infectious particles is observed during prion disease. This can be explained by taking into account fibril breakage. A mathematical solution for the exponential growth rate resulting from the combination of fibril growth and fibril breakage has been found.

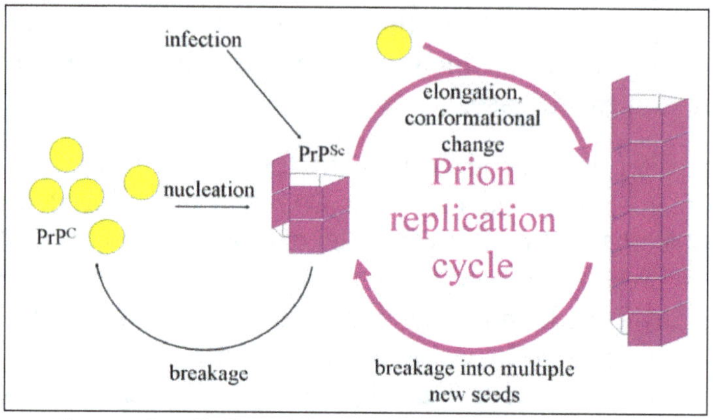

Figure: The Fibril Model: Illustrating prion propagation.

The protein-only hypothesis has been criticized by those who feel that the simplest explanation of the evidence to date is viral.

The virion hypothesis states that TSEs are caused by a replicable informational molecule (likely to be a nucleic acid) bound to PrP. Many TSEs, including scrapie and BSE, show strains with specific and distinct biological properties, a feature which supporters of the virion hypothesis feel is not explained by prions.

Recent studies propagating TSE infectivity in cell-free reactions and in purified component chemical reactions strongly suggest against TSE's viral nature. Using a similar defined recipe of multiple components (PrP, POPG lipid, RNA), Jiyan Ma and colleagues generated infectious prions from recombinant PrP expressed from E. coli, casting further doubt on this hypothesis.

Lichens

Figure: Lichen, Xanthoparmelia cf. lavicola, a foliose lichen on basalt.

Lichen is any of about 15,000 species of plantlike organisms that consist of a symbiotic association of algae (usually green) or cyanobacteria and fungi (mostly ascomycetes and basidiomycetes). Lichens are found worldwide and occur in a variety of environmental conditions. A diverse group of organisms, they can colonize a wide range of surfaces and are frequently found on tree bark, exposed rock, and as a part of biological soil crust. Lichens have been used by humans as food and as sources of medicine and dye. They also provide two-thirds of the food supply for the caribou and reindeer that roam the far northern ranges.

Lichens were once classified as single organisms—until the advent of microscopy, when the association of fungi with algae or cyanobacteria became evident. Although lichens had been assumed to consist of a single fungus species (usually an ascomycete) and a single photosynthetic partner, research suggests that many macrolichens also feature specific basidiomycete yeasts in the cortex of the organism. There is still some discussion about how to classify lichens, though many taxonomists rely on genetic analyses in addition to traditional morphological data.

Figure: Reindeer lichen, Reindeer lichen (Cladonia rangiferina), a type of fruticose lichen.

The composite body of lichen is called a thallus (plural thalli); the body is anchored to its substrate by hair like growths called rhizines. There are three main lichen body types: crustose, fruticose, and foliose. Lichens that form a crustlike covering that is thin and tightly bound to the substrate are called crustose. Squamulose lichens are small and leafy with loose attachments to the substrate and are usually considered to be a special type of crustose lichen. Foliose lichens are large and leafy, reaching diameters of several feet in some species, and are usually attached to the substrate by their large plate like thalli at the centre. These lichens have a distinct top and bottom side and can be leafy, flat, or bumpy and convoluted. Fruticose lichens can be hanging or upright and may be hair like, cuplike, or shrubby in appearance.

In addition to their morphological forms, lichen thalli are also classified by the ratio of phycobiont cells (i.e., cells of the photosynthetic partner) to mycobiont cells (i.e., cells of the fungus). The homoeomerous type of thallus consists of numerous algal cells distributed among a lesser number of fungal cells, while the heteromerous thallus has a predominance of fungal cells.

Evolutionarily, it is not certain when fungi and algae came together to form lichens for the first time, but it was certainly after the mature development of the separate components. As symbionts, the basis of their relationship is the mutual benefit that they provide each other. The photosynthetic algae or cyanobacteria form simple carbohydrates that, when excreted, are absorbed by fungi cells and transformed into a different carbohydrate. In at least one case, Peltigera polydactyla, the exchange occurs within two minutes. The phycobionts also produce vitamins that the fungi need. Fungi contribute to the symbiosis by absorbing water vapour from the air and by providing much-needed shade for the light-sensitive algae beneath.

Figure: Parmelia sulcate. Parmelia sulcata growing on tree bark is a type of foliose lichen.

Lichens are long-lived and grow relatively slowly, and there is still some question as to how they propagate. Most botanists agree that the most common means of reproduction is vegetative; that is, portions of an existing lichen break off and fall away to begin new growth nearby.

Figure: British soldiers lichen. British soldiers lichen (*Cladonia cristatella*), so named for its red fruiting bodies. It is fruticose lichen.

Slime Molds

Slime moulds (or slime molds) are earlier grouped under fungi; however, later they are kept in the kingdom Protista with other unicellular and small multicellular eukaryotic organisms.

They are saprophytic and feed on dead and decaying organic matter. The name 'slime' comes from the gelatinous appearance of macroscopic slime moulds. Under unfavourable conditions, they form aggregates; this is common in plasmodial or acellular slime moulds. Their size varies from a few centimeters to several square metres. They can live as a single-celled organism when there is abundant food, mainly cellular slime moulds.

General Characteristics of Slime Moulds

- Slime moulds are found creeping on debris, decaying leaves or twigs, in soil, on the forest floor, on tree canopies and moist, dark and cool conditions.
- The protoplast is not surrounded by a cell wall in the vegetative phase.
- They are saprophytic and lack chlorophyll. They feed on microorganisms such as bacteria, fungi and yeasts and decompose dead organic matter.
- Some of the slime moulds are parasitic and found in the roots of cabbage and other plants of Brassicaceae family.
- The plasmodial stage resembles protozoa and fruiting bodies form spores resembling fungi.
- Spores have a cell wall made up of cellulose and are resistant to adverse conditions. They can survive for many years.

Slime Moulds Classification

Slime moulds are classified under kingdom Protista. They resemble fungi as well as protozoa. In modern taxonomy, the true slime moulds come under Mycetozoa. They are further classified in different classes. The main classes of slime moulds are the following:

- Myxomycetes: True slime moulds or acellular slime moulds. They are characterised by syncytial (multinucleated), plasmodial stage.
- Dictyostelia: Cellular slime moulds. They do not form huge coenocytes.
- Protostelia: Simple, minute, amoeboid slime moulds.
- Acrasi: Cellular slime moulds similar to dictyostelids but have eruptive pseudopodia.
- Plasmodiophomycetes: Parasitic slime moulds. They are found as an internal parasite in cabbage roots. They cause various diseases in plants such as clubroot disease of cabbage.
- Labyrinthulae: Net slime moulds form a network of tubes in which amoeba without pseudopodia can swim freely.
- Fonticula: Form volcano-shaped fruiting bodies.

Slime moulds are mainly of two types:

- Acellular slime moulds.
- Cellular slime moulds.

Acellular slime moulds are also known as plasmodial slime moulds.

Acellular Slime Moulds (Plasmodial):

- The feeding stage is a multinucleate mass of protoplasm, i.e. plasmodium.

- They can grow up to 1 ft. in diameter.
- They are found creeping as a slimy mass over leaf litter, moist and decaying logs. It feeds on dead and decaying organic matter and microorganisms.
- When the food is scarce and moisture is less, they reproduce asexually.
- Examples: Physarum, Cribaria, Lycogala, Fuligo, Tubifera.

The life cycle of acellular slime moulds:

- Drying plasmodium forms stalked fruiting bodies, which bear sporangia.
- Haploid spores are produced in sporangia by meiosis. They have a thick cell wall and are extremely resistant to adverse conditions.
- Under favourable conditions, these spores germinate to form haploid cells.
- These haploid cells act as a gamete. Different kinds of cells are formed depending on moisture availability.
- Biflagellated swarm cells are formed when the conditions are wet or otherwise amoeboid myxamoebae are formed. These cells act as gametes.
- These haploid gametes fuse to form a diploid zygote.
- The diploid zygote undergoes multiple mitotic divisions forming multinucleated plasmodium as it doesn't undergo cytoplasmic division (cytokinesis).

Cellular Slime Moulds:

- The feeding stage is a single-celled amoeboid, which lives as a solitary organism.
- They have a close resemblance to amoebas.
- Individual cells feed on microorganisms and other food matter while creeping on decaying log or freely swimming in freshwater.
- Just like an amoeba, each cell has a haploid nucleus and divides mitotically.
- When the food is less they form aggregate but retain their individuality due to the presence of a thin plasma membrane and reproduce asexually by spore formation.
- When the food or moisture is depleted, they send out a cAMP-mediated chemical signal.
- The chemical diffuses out and binds to the receptors present on the surface of nearby cells resulting in the movement of cells towards cAMP (cyclic adenosine monophosphate).
- Examples: Dictyostelium, Acytostelium, Polysphondylium.

The lifecycle of cellular slime moulds:

- Thousands of cells aggregate forming a multicellular slug, which creeps for a short distance and settles.

- Each cell in the slug retains its individuality and retains its plasma membrane.
- After settling down, fruiting bodies are formed in the slug, which contains spores.
- Spores are released and germinate under favourable conditions into a single haploid amoeboid cell, which is the vegetative stage.
- Formation of spore is an asexual method of reproduction.
- They mostly lack sexual reproduction and lack the flagellated gamete stage.

References

- Micro, microbiology, Highveld: caister.com, Retrieved 20, March 2020
- Immunology, gram-positive-vs-gram-negative-323007: technologynetworks.com, Retrieved 11, April 2020
- Protozoa: biologydictionary.net, Retrieved 16, June 2020
- Algae-definition-characteristics-and-structure-with-diagram-46727: biologydiscussion.com, Retrieved 01, May 2020
- What-is-microbiology, why-microbiology-matters, viruses: microbiologysociety.org, Retrieved 14, July 2020

Chapter 2

Microbial Metabolism

Microbial metabolism is the means through which microbes obtain energy and nutrients with the help of chemical reactions, in order to maintain life. Microbial catabolism and anabolism are parts of microbial metabolism, each responsible for the breakdown and buildup of complex molecules. This chapter provides a detailed examination of microbial metabolism, its types and its distinct process.

Microbial growth requires the polymerization of biochemical building blocks into proteins, nucleic acids, polysaccharides, and lipids. The building blocks must come preformed in the growth medium or must be synthesized by the growing cells. Additional biosynthetic demands are placed by the requirement for coenzymes that participate in enzymatic catalysis. Biosynthetic polymerization reactions demand the transfer of anhydride bonds from adenosine triphosphate (ATP). Growth demands a source of metabolic energy for the synthesis of anhydride bonds and for the maintenance of trans-membrane gradients of ions and metabolites.

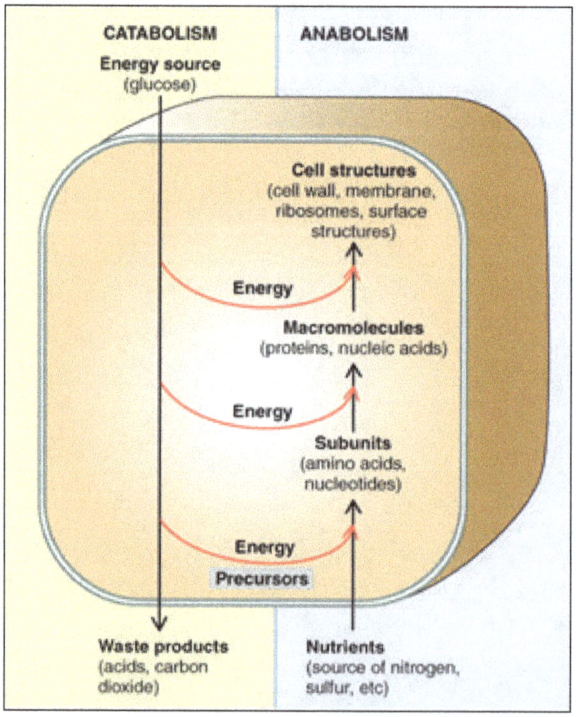

Figure: The relationship between catabolism and anabolism.

Metabolism has two components, catabolism and anabolism. Catabolism encompasses processes that harvest energy released from the breakdown of compounds (e.g., glucose), and using that energy to synthesize ATP. In contrast, anabolism, or biosynthesis, includes processes that utilize the energy stored in ATP to synthesize and assemble the subunits, or building blocks, of macromolecules that make up the cell. The sequence of building blocks within a macromolecule is determined in one of two ways. In nucleic acids and proteins, it is template-directed: DNA serves as the template for its own synthesis and for the synthesis of the various types of RNA; messenger RNA serves as the template for the synthesis of proteins. In carbohydrates and lipids, on the other hand, the arrangement of building blocks is determined entirely by enzyme specificities. Once the macromolecules have been synthesized, they self-assemble to form the supramolecular structures of the cell, e.g., ribosomes, membranes, cell wall, flagella, and pili.

Catabolism encompasses processes that harvest energy released during disassembly of compounds, using it to synthesize adenosine triphosphate (ATP); it also provides precursor metabolites used in biosynthesis. Anabolism, or biosynthesis, includes processes that utilize ATP and precursor metabolites to synthesize and assemble subunits of macromolecules that make up the cell.

The rate of macromolecular synthesis and the activity of metabolic pathways must be regulated so that biosynthesis is balanced. All of the components required for macromolecular synthesis must be present for orderly growth, and control must be exerted so that the resources of the cell are not expended on products that do not contribute to growth or survival.

The following points highlight the five major pathways in organisms. The pathways are:

- Glycolysis.
- Pentose Phosphate Pathway.
- Entner-Doudoroff Pathway.
- Tricarboxylic Acid Cycle.
- Glyoxylate Cycle.

Glycolysis

Glycolysis (glyco-sugar of sweet, lysis-breakdown) is the initial phase of metabolism during which the organic molecule glucose and other sugar are partially oxidized to smaller molecules e.g. pyruvate usually with the generation of some ATP and reduced coenzymes. Microorganisms employ several metabolic pathways to catabolize glucose and other sugars.

There are three important routes of glucose conversion to pyruvate such as glycolysis or Embden-Myerhof pathway (BMP) pathway, pentose phosphate pathway, and Entner-Doudoroff pathway. Glycolysis is the most important type of mechanism by which organisms obtain energy from organic compounds in absence of molecular oxygen. As it occurs in the absence of oxygen, therefore, it is also called anaerobic fermentation.

Since living organisms arose in the environment lacking oxygen, anaerobic fermentation was the only method to obtain energy. However, glycolysis or anaerobic fermentation is present in both aerobic and anaerobic organisms.

Higher organisms have retained the glycolytic pathway of degradation i.e. glucose to pyruvic acid as a preparatory pathway for complete aerobic catabolism of glucose. Glycolysis also serves as an emergency mechanism in anaerobic organisms to produce energy in the absence of oxygen.

EMP Pathways

In case of aerobic catabolic carbohydrate metabolism (aerobic respiration), some bacteria such as E. coli, Azotobacter spp., Bacillus eutrophus, etc. exhibit EMP pathway whereas, ED pathway (phosphorylated) is followed by the species of Alcaligenes, Rhizobium, Xanthomonas, etc. The non-phosphorylated ED pathway occurs in archaea (Pyrococcus spp., Thermoplasma spp, etc.). It is interesting to note that no archaeobacteria uses EMP pathway.

EMP pathway in bacteria initiates by using the phosphoenol pyruvate phosphotransferase system (PEP: PTS) that converts glucose to glucose 6-phosphate during nutrient transport across the cell membrane.

The glucose 6-phosphate is then isomerized to fructose 6-phosphate which is further converted to fructose 1, 6-bi-phosphate. This conversion requires ATP as a source of energy and an enzyme called phosphofructokinase.

It is essentially the reversal of glycolysis, which fulfills a similar anaplerotic role. It is particularly important during growth on pyruvate related C_3 compounds and C_2 units. The several class of flow of carbon from pyruvate maintains a supply of hexoses. These are required for cell wall and its component synthesis.

The complete pathway of glycolysis from glucose to pyruvate were elucidated by Gustav Embden (who gave the manner of cleavage of fructose 1, 6-diphosphate and pattern of subsequent steps) and Otto Meyerhof (who confirmed Embden's work and studied the energetics of glycolysis), in late 1953s. Therefore the sequence reaction from glucose to pyruvate is also called Embden-Meyerhof pathway or glycolysis (EMP).

The overall balance sheet of glycolysis is given below:

$$Glucose + 2ADP + 2Pi + 2NAD^{+} \rightarrow Pyruvate + 2ATP + 2NADH + 2H^{+}$$

In anaerobic organism pyruvate is further converted to lactate or other organic compounds like alcohol, etc., after using NADH and H+ formed during glycolysis:

$$Pyruvate + NADH + H+ \leftrightarrow Lactate + NAD^{+}$$

In aerobes the pyruvate is converted to acetyl CoA as a preparatory step for entrance into tricarboxylic acid cycle, for complete oxidation of glucose.

$$Pyruvate + NAD^{+} + CoA \rightarrow Acetyl\ CoA + NADH + H^{+} + CO_2$$

Glycolysis is carried out by the help of ten enzymes for ten reactions of the glycolytic pathway. These enzymes are present in the soluble portion of the cytoplasm. All the intermediates of the glycolytic pathway are phosphorylated compounds. The most important use of phosphate groups is in the production of ATP from ADP and phosphate.

The complete reactions of glycolytic pathway can be divided into two stages. In the first stage, ATP is utilized and glucose is converted into two molecules of three carbon compounds, glyceraldehyde

3-phosphate and dihydroxy acetone phosphate. The glyceraldehyde 3-phosphate is converted into pyruvic acid resulting in a net synthesis of two molecules of ATP. The complete reaction with respective enzyme is shown below.

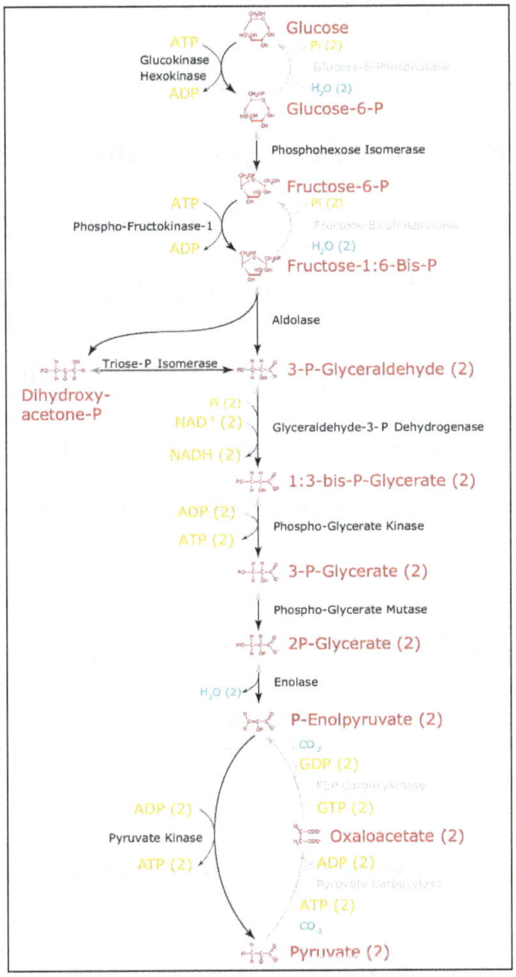

Figure: Glycolytic pathway showing synthesis of two ATP molecules and pyruvic acid (glycolysis).

Apart from glucose, other types of sugar (monosaccharides, disaccharides, polysaccharides) can also enter the glycolytic pathway.

Polysaccharides e.g. Glycogen

$$(Glucose)n + HPO_4^{2-} \xrightarrow[Glycogen]{Glycogenphosphorylase} (Glucose)_{n-1} + Glucose-1-phosphate$$

$$Glucose-1-phosphate \xrightarrow{Phosphoglucomutase} Glucose-6-phosphate$$

Disaccharides: e.g. Sucrose

$$Sucrose + H_2O \xrightarrow{Fructuranosidase} Glucose + Fructose$$

$$Maltose + H_2O \xrightarrow{Glucosidase} 2Glucose$$

$$Lactose + H_2O \xrightarrow{Galactosidase} Glucose + Galactose$$

The three key regulatory enzymes, hexokinase, phosphofructokinase and pyruvate kinase act irreversibly and rest of the steps are reversible.

Homo-saccharides: e.g. Fructose

Fructose can enter the glycolysis by changing to glyceraldehyde 3-phosphate.

$$Fructose + ATP \xrightarrow{Fructokinase} Fructose\ 1 - Phosphate + ADP$$

$$Fructose - 1\ phosphate \xrightarrow{Aldolase} Glyceraldehyde + dihydroxyacetone\ phosphate$$

$$Glyceraldehyde + ATP \longrightarrow Glyceraldehyde - 3 - phosphate + ADP$$

Mannose enters glycolysis after converting to fructose 6-phosphate.

$$Mannose + ATP \xrightarrow{Hexokinase} Mannose\ 6 - phosphate$$

Glycerol can also enter the glycolysis.

$$Glycerol + ATP \longrightarrow Glycerol\ 3 - phosphate + ADP$$

$$Glycerol\ 3 - phospahte + NAD^+ \xrightarrow[dehydrogenase]{Glycerol\ 3-phosphate} Dihydroxyacetone\ phosphate + NADH + H^+$$

Dihydroxyacetone phosphate can enter the glycolysis after enzymatically converting to dihydroxyacetone phosphate.

Alternate EMP Pathway-Methyl Glyoxal Pathway

The methyl glyoxal pathway is an alternate of the EMP pathway. It Operates in the presence of low concentration of phosphate to the bacteria, E. coli, Clostridium spp., Pseudomonas spp. etc. In this pathway, dihydroxyacetone so formed converted to methyl glyoxal which later on gives rise to pyruvate.

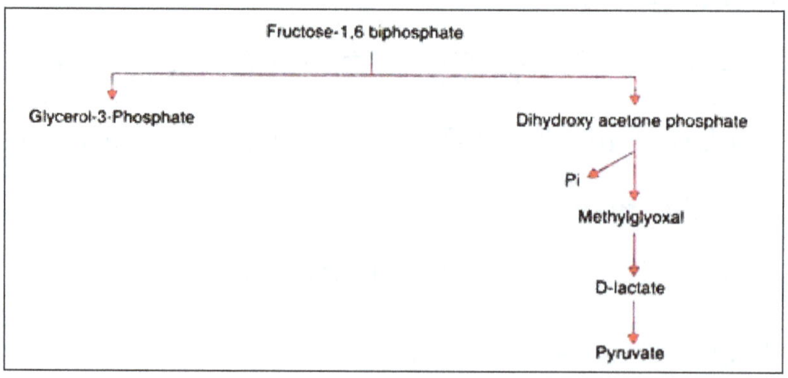

Hence, there is complete absence of the phosphorylation step in which glyceraldehyde 3-phosphate forms 1, 3-bis-phospho- glycerate. The methyl glyoxal pathway consumes O_2 and ATP and no ATP is generated in this pathway.

Pentose Phosphate Pathway (PPP)

Pentose phosphate pathway is an alternative of glucose degradation. This pathway, also called hexose monophosphate shunt (HMP) or phosphogluconate pathway is not the major pathway, but is a multipurpose pathway. Its main function is to generate reducing power in the extra-mitochondrial cytoplasm in the form of NADH. Its second function is to convert hexoses into pentoses, required in synthesis of nucleic acids.

Its third function is complete oxidative degradation of pentose. The reactions of phosphogluconate pathway take place in the soluble portion of extra-mitochondrial cytoplasm of cells.

The bacteria which show PPP are Bacillus subtilis, E. coli. Streptococcus faecalis and Leuconostoc mesenteroides. Apart from microorganisms the prominent tissues which show PPP are liver, mammary gland and adrenal cortex.

There are three enzymes involved in PPP i.e. transketolase, transaldolase and ribulosephosphate 3-epimerase. Ribulose phosphate 3-epimerase catalyzes the conversion of ribulose 5-phosphate into the epimer xylulose 5-phosphate. Transketolase transfers the glycoaldehyde group (CH, OH—CO—) from xylulose 5-phosphate to ribose 5-phosphate to yield sedoheptulose 7-phosphate and glyceraldehyde-3-phosphate.

Transketolase also catalyzes the transfer of glycoaldehyde group from a number of 2-keto sugar phosphate to carbon atom one of a number of different aldose phosphate. Transaldolase acts on the products of Transketolase and transfer dihydroxyacetone group to form fructose 6-phosphate and erythrose 4-phosphate.

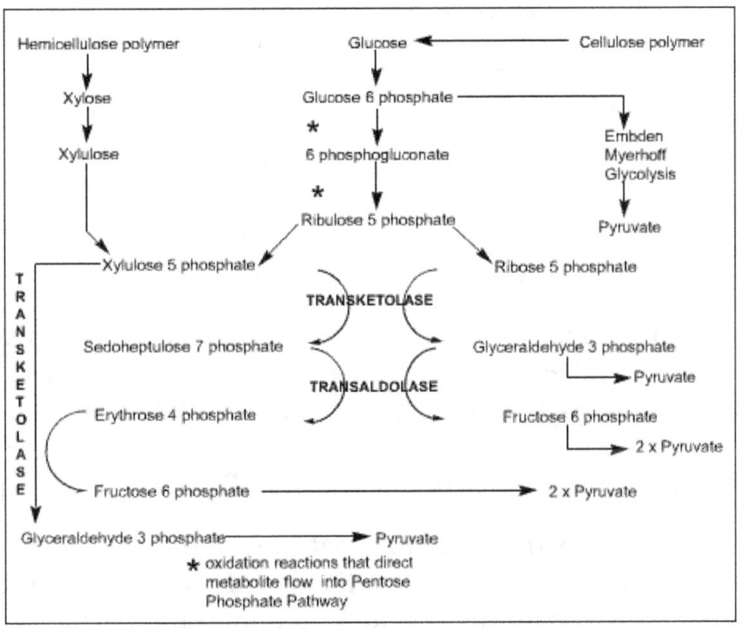

Figure: The Pentose phosphate pathway.

Pentose phosphate pathway thus functions according to the needs of the cell. If the requirement of reducing power is more then it proceeds towards the formation of NADPH but if pentoses are required it functions in the direction of formation of pentose. But if the cell requires instant energy the PPP stops and glycolysis and TCA proceed.

- To anabolic reactions that require electron donors.
- To Calvin-Benson Cycle (dark reactions of photosynthesis).
- To synthesis of nucleotides and nucleic acids.
- To step e of glycolysis.
- To glucose 6-phosphate which can enter the pentose phosphate pathway or glycolysis.
- To synthesis of several amino acids.

Entner-Doudoroff Pathway

Apart from glycolysis, Entner-Doudoroff pathway is another pathway for oxidation of glucose to pyruvic acid. This pathway is found in some Gram-negative bacteria like Rhizobium, Agrobacterium and Pseudomonas and is absent in Gram-positive bacteria. In this pathway each molecule of glucose, forms two molecules of NADPH and one molecule of ATP. The complete pathway is shown in the figure below.

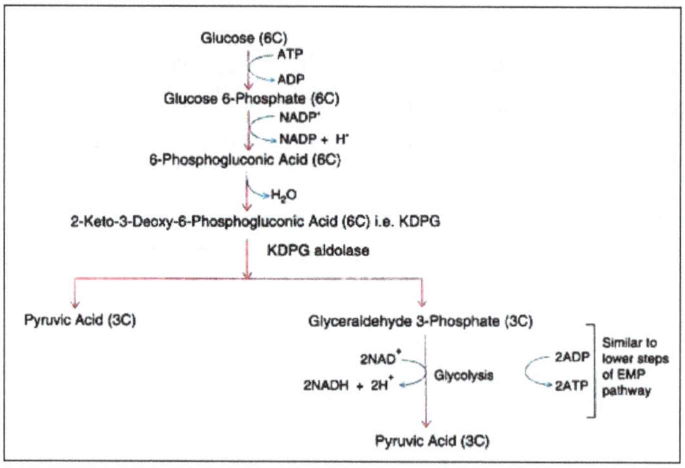

Figure: Entner-Doudoroff pathway.

In this pathway glucose 6-phosphate is oxidized to 6-phosphogluconate, then converted to 2- keto-3-deoxy-6-phosphoglucose (kDPG) cleaved using enzyme to give rise glyceraldehyde's 3- phosphate and pyruvate directly without generation of ATP. The catabolism of glucose results in production of only one ATP molecule whereas in EMP pathway, two ATP molecules are produced. This seems that EMP pathway more efficient than that of ED pathway.

Further, difference between ED pathway and PP pathway is the generation of reduced NADPH from NADP in the former. It is interesting to note that coenzyme NADP+ and NADPH are used in anabolic reactions. Thus, the ED pathway provides an important mechanism for producing NADPH and the 3-carbon building blocks used in biosynthetic reactions etc.

Partially non-phosphorylated ED pathway is involved in some bacteria such as Clostridium spp. Achromobacter spp., Alcaligens spp. and Archaea (Halobacterium spp.) In this case, intermediate product formed prior to kDPG is non-phosphorylated, and phosphogluconate is dehydrated to give rise kDPG, which gives to pyruvate.

In later steps, the reactions of ED pathway are followed. This pathway is also found in other bacteria such as Pseudomonas aeruginosa, Azotobacter, and Enterococcus faecalis, and an anaerobic bacterium Zymomonas mobilis.

Tricarboxylic Acid Cycle

The tricarboxylic acid cycle was first given by H.A. Krebs in 1937. H.A. Krebs then gave the name citric acid cycle. Because of citric acid is the first product of Krebs cycle, is also known as TCA cycle due to presence of three carboxylic groups in a molecule of citric acid. The cycle is of universal occurrence in all the aerobic organisms and leads to complete oxidation of glucose to CO_2 and H_2O while glycolysis leads to incomplete oxidation of glucose to pyruvate.

Tri-carboxyhc acid cycle completely oxidizes it to release large amount of energy in the form of NADH + H^+ mainly and GTP. NADH + H^+ enter the respiratory chain where each NADH + H^+ produces three ATP molecules. GTP is converted to ATP by substrate level oxidation. Another form of energy is in the form of substrate of $FADH^2$, which also enters the respiratory chain to form two molecules of ATP.

All the reactions of tricarboxylic acid cycle take place in the inner compartment of mitochondrion. Some of these enzymes occur in the matrix of inner compartment, while rest of them occurs on the inner mitochondrial membrane. For the start of the cycle, the pyruvate formed in the glycolysis is first converted to acetyl Co-A by preparatory reaction.

$$Pyruvate + NAD^+ + CoA \rightarrow acetyl\ CoA + NADH + H+ + CO_2$$

The reaction is irreversible and is not itself a part of the tricarboxylic acid cycle. It is carried out with the help of the enzyme pyruvate dehydrogenase. Acetyl CoA then enters the cycle after combining with oxaloacetate to form citrate, after which a cycle of reactions occurs leading to the formation of six CO_2, eight NADH + H^+, one $FADH_2$ and one molecule of glucose.

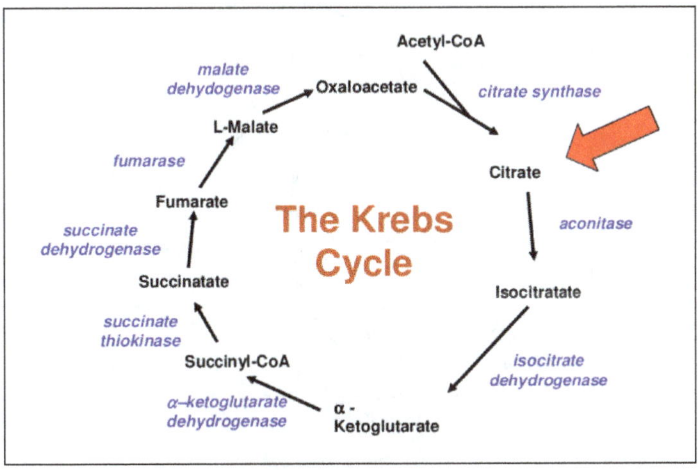

Figure: Kreb's cycle.

There are few key steps in the tricarboxylic acid cycle which control the cycle as per need of the cell. The first of these controls is the preparatory reaction. The activity of pyruvate dehydrogenase is reduced in the presence of excess ATP and again increases in the absence of ATP.

There are two more steps which can control the cycle. These are the isocitrate dehydrogenase reaction (which requires ADP as positive regulation), and succinate dehydrogenase reaction (promoted by succinate, phosphate and ATP). However, the key control of the cycle is the reaction carried out by citrate synthase. This is the primary control of the cycle.

Glyoxylate Cycle

It is anaplerotic reaction in which oxaloacetate is taken from TCA cycle to meet out the demand of carbon requirement for amino acid biosynthesis. Hence, these intermediates have to be replenished via an alternate route, called anaplerotic pathway i.e. glyoxylate pathway. This cycle operates for glucoiieogenesis. Glyoxylate cycle was given first by Krebs and H.R. Kornberg.

This cycle is a modified form of tricarboxylic acid cycle found in plants and those microorganisms which utilize fatty acids as the source of energy in the form of acetyl Co A.

In this cycle the CO_2 evolving steps of tricarboxylic acid cycle were by-passed and instead a second molecule of acetyl CoA is utilized (which condenses with glyoxylate to form malate). Succinate is a byproduct, used for biosynthesis, particularly in gluconeogenesis. The overall reaction of glyoxylate cycle is given below:

$$2\,Acetyl\ Co - A\ +\ NAD^{+} + 2H_2O\ \rightarrow\ Succinate\ +\ 2CoA\ +\ NADH\ +\ H^{+}$$

The two key enzymes, isocitrate lyase and malate synthase of glyoxylate cycle are localized in cytoplasmic organelles called glyoxysomes. Glyoxylate cycle goes on simultaneously with the tricarboxylic acid cycle, while tricarboxylic acid provides energy; glyoxylate cycle provides succinate for the formation of new carbohydrate from fats as shown in Figure below.

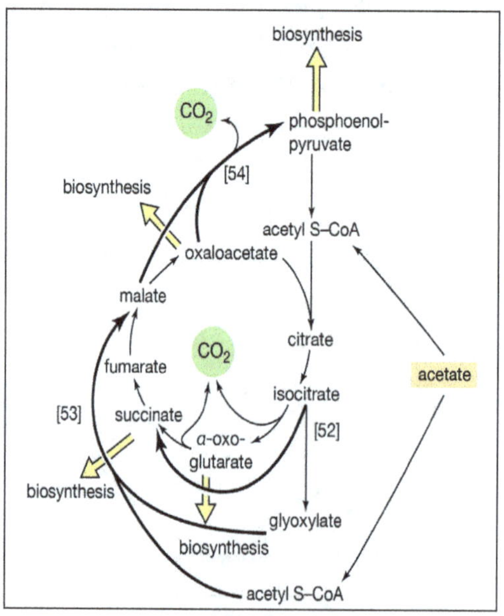

Figure: Glyoxylate cycle.

Energy, Matter and Enzymes

The term used to describe all of the chemical reactions inside a cell is metabolism. Cellular processes such as the building or breaking down of complex molecules occur through series of stepwise, interconnected chemical reactions called metabolic pathways. Reactions that are spontaneous and release energy are exergonic reactions, whereas endergonic reactions require energy to proceed. The term anabolism refers to those endergonic metabolic pathways involved in biosynthesis, converting simple molecular building blocks into more complex molecules, and fueled by the use of cellular energy. Conversely, the term catabolism refers to exergonic pathways that break down complex molecules into simpler ones. Molecular energy stored in the bonds of complex molecules is released in catabolic pathways and harvested in such a way that it can be used to produce high-energy molecules, which are used to drive anabolic pathways. Thus, in terms of energy and molecules, cells are continually balancing catabolism with anabolism.

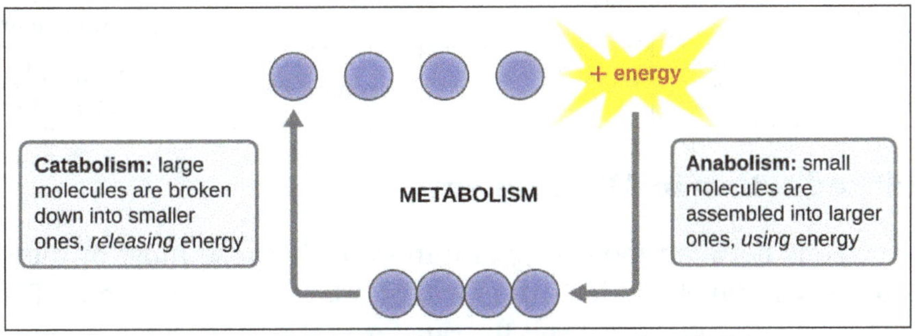

Figure: Metabolism includes catabolism and anabolism.

Anabolic pathways require energy to synthesize larger molecules. Catabolic pathways generate energy by breaking down larger molecules. Both types of pathways are required for maintaining the cell's energy balance.

Classification by Carbon and Energy Source

Organisms can be identified according to the source of carbon they use for metabolism as well as their energy source. The prefixes auto- ("self") and hetero- ("other") refer to the origins of the carbon sources various organisms can use. Organisms that convert inorganic carbon dioxide (CO_2) into organic carbon compounds are autotrophs. Plants and cyanobacteria are well-known examples of autotrophs. Conversely, heterotrophs rely on more complex organic carbon compounds as nutrients; these are provided to them initially by autotrophs. Many organisms, ranging from humans to many prokaryotes, including the well-studied Escherichia coli, are heterotrophic.

Organisms can also be identified by the energy source they use. All energy is derived from the transfer of electrons, but the source of electrons differs between various types of organisms. The prefixes photo- ("light") and chemo- ("chemical") refer to the energy sources that various organisms use. Those that get their energy for electron transfer from light are phototrophs, whereas chemotrophs obtain energy for electron transfer by breaking chemical bonds. There are two types of chemotrophs: organotrophs and lithotrophs. Organotrophs, including humans, fungi, and many prokaryotes, are chemotrophs that obtain energy from organic compounds. Lithotrophs ("litho" means "rock") are chemotrophs that get energy from inorganic compounds, including hydrogen sulfide (H_2S) and reduced iron. Lithotrophy is unique to the microbial world.

The strategies used to obtain both carbon and energy can be combined for the classification of organisms according to nutritional type. Most organisms are chemoheterotrophs because they use organic molecules as both their electron and carbon sources.

Table: Classifications of Organisms by Energy and Carbon source.

Classifications		Energy Source	Carbon Source	Examples
Chemotrophs	Chemoautotrophs	Chemical	Inorganic	Hydrogen-, sulfur-, iron-, nitrogen-, and carbon monoxide-oxidizing bacteria
	Chemoheterotrophs	Chemical	Organic compounds	All animals, most fungi, protozoa, and bacteria
Phototrophs	Photoautotrophs	Light	Inorganic	All plants, algae, cyanobacteria, and green and purple sulfur bacteria
	Photoautotrophs	Light	Organic compounds	Green and purple non-sulfur bacteria, heliobacteria

Oxidation and Reduction in Metabolism

The transfer of electrons between molecules is important because most of the energy stored in atoms and used to fuel cell functions is in the form of high-energy electrons. The transfer of energy in the form of electrons allows the cell to transfer and use energy incrementally; that is, in small packages rather than a single, destructive burst. Reactions that remove electrons from donor molecules, leaving them oxidized are oxidation reactions; those that add electrons to acceptor molecules, leaving them reduced, are reduction reactions. Because electrons can move from one molecule to another, oxidation and reduction occur in tandem. These pairs of reactions are called oxidation-reduction reactions, or redox reactions.

Energy Carriers: NAD+, NADP+, FAD and ATP

The energy released from the breakdown of the chemical bonds within nutrients can be stored either through the reduction of electron carriers or in the bonds of adenosine triphosphate (ATP). In living systems, a small class of compounds functions as mobile electron carriers, molecules that bind to and shuttle high-energy electrons between compounds in pathways. The principal electron carriers we will consider originate from the B vitamin group and are derivatives of nucleotides; they are nicotinamide adenine dinucleotide, nicotine adenine dinucleotide phosphate, and flavin adenine dinucleotide. These compounds can be easily reduced or oxidized. Nicotinamide adenine dinucleotide (NAD+/NADH) is the most common mobile electron carrier used in catabolism. NAD+ is the oxidized form of the molecule; NADH is the reduced form of the molecule. Nicotine adenine dinucleotide phosphate (NADP+), the oxidized form of an NAD+ variant that contains an extra phosphate group, is another important electron carrier; it forms NADPH when reduced. The oxidized form of flavin adenine dinucleotide is FAD, and its reduced form is FADH2. Both NAD+/NADH and FAD/FADH2 are extensively used in energy extraction from sugars during catabolism in chemoheterotrophs, whereas NADP+/NADPH plays an important role in anabolic reactions and photosynthesis. Collectively, FADH2, NADH, and NADPH are often referred to as having reducing power due to their ability to donate electrons to various chemical reactions.

A living cell must be able to handle the energy released during catabolism in a way that enables the cell to store energy safely and release it for use only as needed. Living cells accomplish this by using the compound adenosine triphosphate (ATP). ATP is often called the "energy currency" of the cell, and, like currency, this versatile compound can be used to fill any energy need of the cell. At the heart of ATP is a molecule of adenosine monophosphate (AMP), which is composed of an adenine molecule bonded to a ribose molecule and a single phosphate group. Ribose is a five-carbon sugar found in RNA, and AMP is one of the nucleotides in RNA. The addition of a second phosphate group to this core molecule results in the formation of adenosine diphosphate (ADP); the addition of a third phosphate group forms ATP.

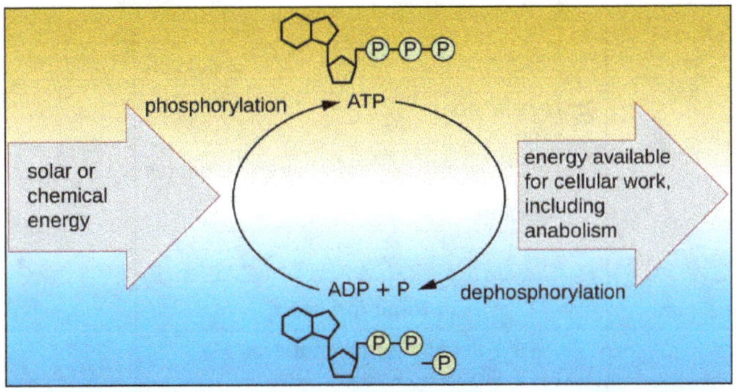

Figure: The energy released from dephosphorylation of ATP is used to drive cellular work, including anabolic pathways. ATP is regenerated through phosphorylation, harnessing the energy found in chemicals or from sunlight.

Adding a phosphate group to a molecule, a process called phosphorylation, requires energy. Phosphate groups are negatively charged and thus repel one another when they are arranged in series, as they are in ADP and ATP. This repulsion makes the ADP and ATP molecules inherently unstable. Thus, the bonds between phosphate groups (one in ADP and two in ATP) are called high-energy phosphate bonds. When these high-energy bonds are broken to release one phosphate (called inorganic phosphate [P_i]) or two connected phosphate groups (called pyrophosphate [PP_i]) from ATP through a process called dephosphorylation, energy is released to drive endergonic reactions.

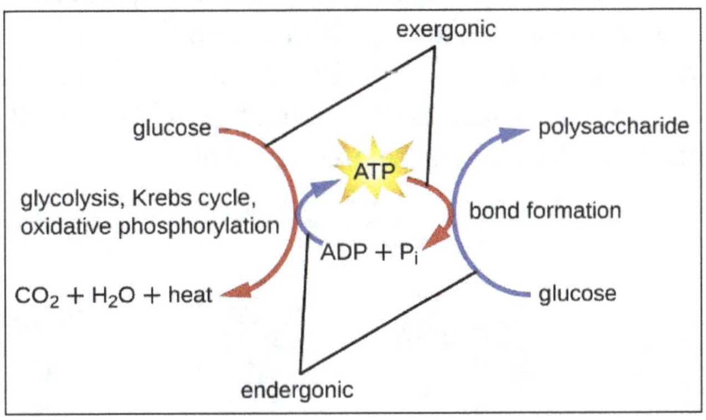

Figure: Exergonic reactions are coupled to endergonic ones, making the combination favorable.

Here, the endergonic reaction of ATP phosphorylation is coupled to the exergonic reactions of catabolism. Similarly, the exergonic reaction of ATP dephosphorylation is coupled to the endergonic reaction of polypeptide formation, an example of anabolism.

Enzyme Structure and Function

A substance that helps speed up a chemical reaction is a catalyst. Catalysts are not used or changed during chemical reactions and, therefore, are reusable. Whereas inorganic molecules may serve as catalysts for a wide range of chemical reactions, proteins called enzymes serve as catalysts for biochemical reactions inside cells. Enzymes thus play an important role in controlling cellular metabolism.

An enzyme functions by lowering the activation energy of a chemical reaction inside the cell. Activation energy is the energy needed to form or break chemical bonds and convert reactants to products. Enzymes lower the activation energy by binding to the reactant molecules and holding them in such a way as to speed up the reaction.

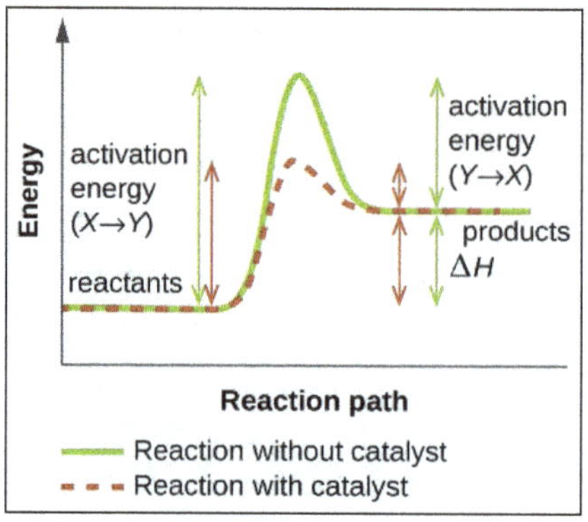

Figure: Enzymes lower the activation energy of a chemical reaction.

The chemical reactants to which an enzyme binds are called substrates, and the location within the enzyme where the substrate binds is called the enzyme's active site. The characteristics of the amino acids near the active site create a very specific chemical environment within the active site that induces suitability to binding, albeit briefly, to a specific substrate (or substrates). Due to this jigsaw puzzle-like match between an enzyme and its substrates, enzymes are known for their specificity. In fact, as an enzyme binds to its substrate(s), the enzyme structure changes slightly to find the best fit between the transition state (a structural intermediate between the substrate and product) and the active site, just as a rubber glove molds to a hand inserted into it. This active-site modification in the presence of substrate, along with the simultaneous formation of the transition state, is called induced fit. Overall, there is a specifically matched enzyme for each substrate and, thus, for each chemical reaction; however, there is some flexibility as well. Some enzymes have the ability to act on several different structurally related substrates.

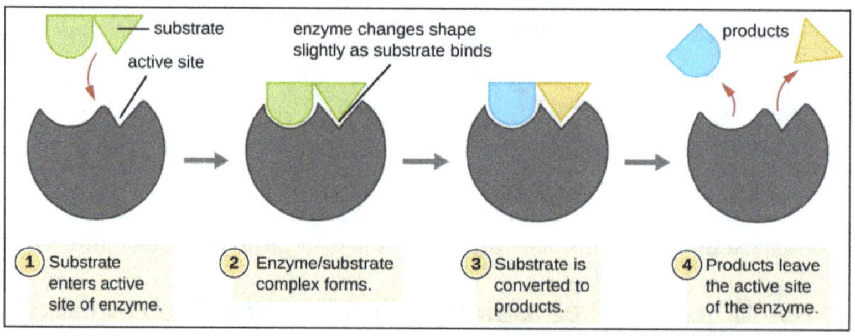

Figure: According to the induced-fit model, the active site of the enzyme undergoes conformational changes upon binding with the substrate.

Enzymes are subject to influences by local environmental conditions such as pH, substrate concentration, and temperature. Although increasing the environmental temperature generally increases reaction rates, enzyme catalyzed or otherwise, increasing or decreasing the temperature outside of an optimal range can affect chemical bonds within the active site, making them less well suited to bind substrates. High temperatures will eventually cause enzymes, like other biological molecules, to denature, losing their three-dimensional structure and function. Enzymes are also suited to function best within a certain pH range, and, as with temperature, extreme environmental pH values (acidic or basic) can cause enzymes to denature. Active-site amino-acid side chains have their own acidic or basic properties that are optimal for catalysis and, therefore, are sensitive to changes in pH.

Another factor that influences enzyme activity is substrate concentration: Enzyme activity is increased at higher concentrations of substrate until it reaches a saturation point at which the enzyme can bind no additional substrate. Overall, enzymes are optimized to work best under the environmental conditions in which the organisms that produce them live. For example, while microbes that inhabit hot springs have enzymes that work best at high temperatures, human pathogens have enzymes that work best at 37 °C. Similarly, while enzymes produced by most organisms work best at a neutral pH, microbes growing in acidic environments make enzymes optimized to low pH conditions, allowing for their growth at those conditions.

Many enzymes do not work optimally, or even at all, unless bound to other specific non-protein helper molecules, either temporarily through ionic or hydrogen bonds or permanently through stronger covalent bonds. Binding to these molecules promotes optimal conformation and function for their respective enzymes. Two types of helper molecules are cofactors and coenzymes. Cofactors are inorganic ions such as iron (Fe^{2+}) and magnesium (Mg^{2+}) that help stabilize enzyme conformation and function. One example of an enzyme that requires a metal ion as a cofactor is the enzyme that builds DNA molecules, DNA polymerase, which requires a bound zinc ion (Zn^{2+}) to function.

Coenzymes are organic helper molecules that are required for enzyme action. Like enzymes, they are not consumed and, hence, are reusable. The most common sources of coenzymes are dietary vitamins. Some vitamins are precursors to coenzymes and others act directly as coenzymes.

Some cofactors and coenzymes, like coenzyme A (CoA), often bind to the enzyme's active site, aiding in the chemistry of the transition of a substrate to a product. In such cases, an enzyme lacking a necessary cofactor or coenzyme is called an apoenzyme and is inactive. Conversely, an enzyme with the necessary associated cofactor or coenzyme is called a holoenzyme and is active. NADH and ATP are also both examples of commonly used coenzymes that provide high-energy electrons or phosphate groups, respectively, which bind to enzymes, thereby activating them.

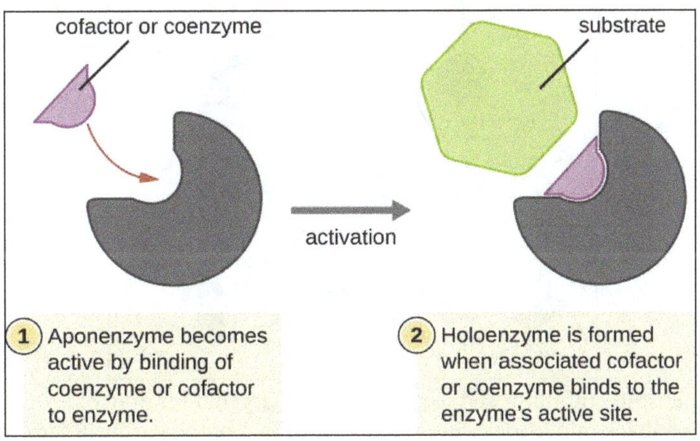

Figure: The binding of a coenzyme or cofactor to an apoenzyme is often required to form an active holoenzyme.

Enzyme Inhibitors

Enzymes can be regulated in ways that either promote or reduce their activity. There are many different kinds of molecules that inhibit or promote enzyme function, and various mechanisms exist for doing so. A competitive inhibitor is a molecule similar enough to a substrate that it can compete with the substrate for binding to the active site by simply blocking the substrate from binding. For a competitive inhibitor to be effective, the inhibitor concentration needs to be approximately equal to the substrate concentration. Sulfa drugs provide a good example of competitive competition. They are used to treat bacterial infections because they bind to the active site of an enzyme within the bacterial folic acid synthesis pathway. When present in a sufficient dose, a sulfa drug prevents folic acid synthesis, and bacteria are unable to grow because they cannot synthesize DNA, RNA, and proteins. Humans are unaffected because we obtain folic acid from our diets.

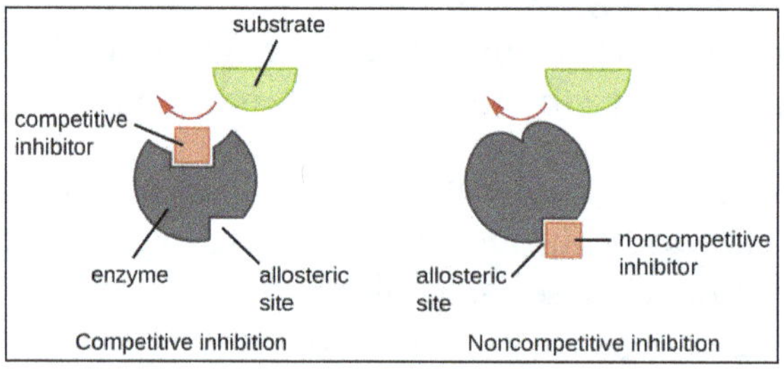

Figure: Enzyme activity can be regulated by either competitive inhibitors, which bind to the active site, or noncompetitive inhibitors, which bind to an allosteric site.

On the other hand, a noncompetitive (allosteric) inhibitor binds to the enzyme at an allosteric site, a location other than the active site, and still manages to block substrate binding to the active site by inducing a conformational change that reduces the affinity of the enzyme for its substrate. Because only one inhibitor molecule is needed per enzyme for effective inhibition, the concentration of inhibitors needed for noncompetitive inhibition is typically much lower than the substrate concentration.

In addition to allosteric inhibitors, there are allosteric activators that bind to locations on an enzyme away from the active site, inducing a conformational change that increases the affinity of the enzyme's active site(s) for its substrate(s).

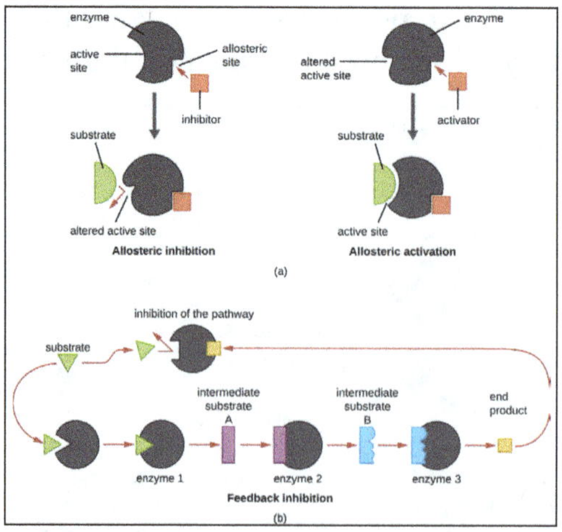

Figure: (a) Binding of an allosteric inhibitor reduces enzyme activity, but binding of an allosteric activator increases enzyme activity. (b) Feedback inhibition, where the end product of the pathway serves as a noncompetitive inhibitor to an enzyme early in the pathway, is an important mechanism of allosteric regulation in cells.

Allosteric control is an important mechanism of regulation of metabolic pathways involved in both catabolism and anabolism. In a most efficient and elegant way, cells have evolved also to use the products of their own metabolic reactions for feedback inhibition of enzyme activity. Feedback inhibition involves the use of a pathway product to regulate its own further production. The cell responds to the abundance of specific products by slowing production during anabolic or catabolic reactions.

Cellular Respiration

Microorganisms such as cyanobacteria can trap the energy in sunlight through the process of photosynthesis and store it in the chemical bonds of carbohydrate molecules. The principal carbohydrate formed in photosynthesis is glucose. Other types of microorganisms such as non-photosynthetic bacteria, fungi, and protozoa are unable to perform this process. Therefore, these organisms must rely upon preformed carbohydrates in the environment to obtain the energy necessary for their metabolic processes.

Cellular respiration is the process by which microorganisms obtain the energy available in carbohydrates. They take the carbohydrates into their cytoplasm, and through a complex series of metabolic processes, they break down the carbohydrate and release the energy. The energy is generally not needed immediately, so it is used to combine ADP with phosphate ions to form ATP molecules. During the process of cellular respiration, carbon dioxide is given off as a waste product. This carbon dioxide can be used by photosynthesizing cells to form new carbohydrates. Also in the process of cellular respiration, oxygen gas is required to serve as an acceptor of electrons. This oxygen gas is identical to the oxygen gas given off in photosynthesis.

The overall mechanism of cellular respiration involves four subdivisions:

- Glycolysis: In which glucose molecules are broken down to form pyruvic acid molecules.
- Krebs cycle: In which pyruvic acid is further broken down and the energy in its molecule is used to form high-energy compounds such as NADH.
- Electron transport system: In which electrons are transported along a series of coenzymes and cytochromes and the energy in the electrons is released.
- Chemiosmosis: In which the energy given off by electrons is used to pump protons across a membrane and provide the energy for ATP synthesis.

Glycolysis

The process of glycolysis is a multistep metabolic pathway that occurs in the cytoplasm of microbial cells and the cells of other organisms. At least six enzymes operate in the metabolic pathway.

In the first and third steps of the pathway, ATP is used to energize the molecules. Thus, two molecules of ATP must be expended in the process. Further along in the process, the six-carbon glucose molecule is converted into intermediary compounds and then is split into two three-carbon compounds. The latter undergo additional conversions and eventually form pyruvic acid at the conclusion of the process.

During the latter stages of glycolysis, four ATP molecules are synthesized using the energy given off during the chemical reactions. Thus, four ATP molecules are synthesized and two ATP molecules are inserted into the process for a net gain of two ATP molecules in glycolysis.

Also during glycolysis, another of the reactions yields enough energy to convert NAD to NADH. The reduced coenzyme (NADH) will later be used in the electron transport system, and its energy will be released. During glycolysis, two NADH molecules are produced.

As glycolysis does not use oxygen, the process is considered to be anaerobic. For certain anaerobic organisms, such as certain bacteria and fermentation yeasts, glycolysis is the sole source of energy. It is a somewhat inefficient process because much of the cellular energy remains in the two molecules of pyruvic acid.

The Krebs Cycle

Following glycolysis, the mechanism of cellular respiration then involves another multistep process called the Krebs cycle, also called the citric acid cycle and the tricarboxylic acid cycle. The Krebs cycle uses the two molecules of pyruvic acid formed in glycolysis and yields high-energy molecules of NADH and FADH and some ATP and carbon dioxide.

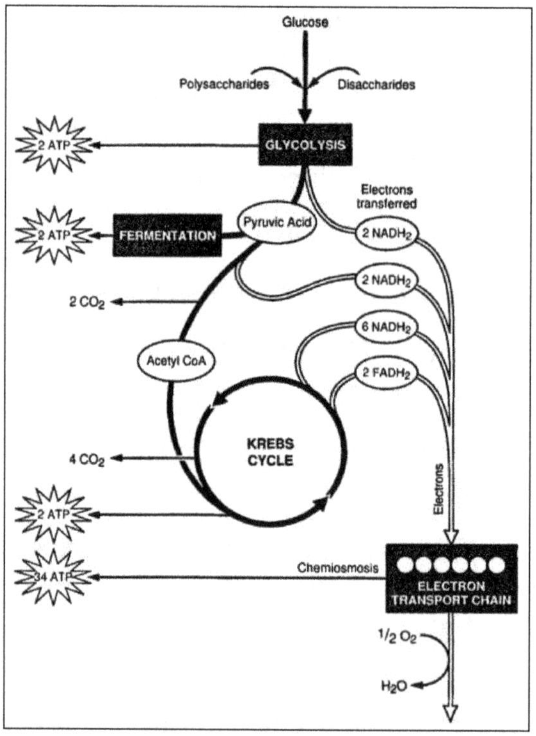

Figure: The processes of cellular respiration showing the major pathways and the places where ATP is synthesized.

The Krebs cycle occurs at the cell membrane of bacterial cells and in the mitochondria of eukaryotic cells. Each of these sausage-shaped organelles of eukaryotic microorganisms possesses inner and outer membranes, and therefore an inner and outer compartment. The inner membrane is folded over itself many times; the folds are called cristae. Along the cristae are the important enzymes necessary for the proton pump and for ATP production.

Prior to entering the Krebs cycle, the pyruvic acid molecules are processed. Each three-carbon molecule of pyruvic acid undergoes conversion to a substance called acetyl-coenzyme A, or acetyl-CoA. In the process, the pyruvic acid molecule is broken down by an enzyme, one carbon atom is released in the form of carbon dioxide, and the remaining two carbon atoms are combined with a coenzyme called coenzyme A. This combination forms acetyl-CoA. In the process, electrons and a hydrogen ion are transferred to NAD to form high-energy NADH.

Acetyl-CoA now enters the Krebs cycle by combining with a four-carbon acid called oxaloacetic acid. The combination forms the six-carbon acid called citric acid. Citric acid undergoes a series of enzyme-catalyzed conversions. The conversions, which involve up to 10 chemical reactions, are all brought about by enzymes. In many of the steps, high-energy electrons are released to NAD. The NAD molecule also acquires a hydrogen ion and becomes NADH. In one of the steps, FAD serves as the electron acceptor, and it acquires two hydrogen ions to become $FADH_2$. Also, in one of the reactions, enough energy is released to synthesize a molecule of ATP. Since there are two pyruvic acid molecules entering the system, two ATP molecules are formed.

Also during the Krebs cycle, the two carbon atoms of acetyl-CoA are released and each forms a carbon dioxide molecule. Thus, for each acetyl-CoA entering the cycle, two carbon dioxide molecules are formed. Since two acetyl-CoA molecules enter the cycle, and each has two carbon atoms, four carbon dioxide molecules will form. Add these four molecules to the two carbon dioxide molecules formed in the conversion of pyruvic acid to acetyl-CoA, and the total is six carbon dioxide molecules. These six CO_2 molecules are given off as waste gas in the Krebs cycle. They represent the six carbons of glucose that originally entered the process of glycolysis.

At the end of the Krebs cycle, the final product formed is oxaloacetic acid, identical to the oxaloacetic acid which begins the cycle. The molecule is now ready to accept another acetyl-CoA molecule to begin another turn of the cycle. All told, the Krebs cycle forms (per two molecules of pyruvic acid) two ATP molecules, a large number of NADH molecules, and some $FADH_2$ molecules. The NADH and the $FADH_2$ will be used in the electron transport system.

The Electron Transport System

The electron transport system occurs at the bacterial cell membrane and in the cristae of the mitochondria in eukaryotic cells. Here, a series of cytochromes (cell pigments) and coenzymes exist. These cytochromes and coenzymes act as carrier molecules and transfer molecules. They accept high-energy electrons and pass the electrons to the next molecule in the system. At key proton-pumping sites, the energy of the electrons is used to transport protons across the cell membrane or into the outer compartment of the mitochondrion.

Each NADH molecule is highly energetic. It accounts for the transfer of six protons across the membrane. Each $FADH_2$ molecule accounts for the transfer of four protons. Electrons pass from NAD to FAD, to other cytochromes and coenzymes, and eventually they lose much of their energy. The final electron acceptor is an oxygen atom. The electron-oxygen combination then takes on two protons to form a molecule of water (H_2O). As a final electron receptor, oxygen is responsible for removing electrons from the system. If oxygen were not available, electrons could not be passed among the coenzymes, the energy in electrons could not be released, the proton pump could not be established, and ATP could not be produced.

Chemiosmosis

The actual production of ATP in cellular respiration takes place during chemiosmosis. As previously noted, chemiosmosis involves the pumping of protons through special channels in the membranes of mitochondria from the inner to the outer compartment. In bacteria, the pumping occurs at the cell membrane. The pumping establishes a proton gradient. Once the gradient is established, protons pass down the gradient through molecular particles. In these particles, the energy of the protons is used to generate ATP, using ADP and phosphate ions as the starting points.

The energy production in cellular respiration during chemiosmosis is substantial. Most biochemists agree that in prokaryotic microorganisms, a total of 36 molecules of ATP can be produced during cellular respiration. In eukaryotic cells, the number is 34 molecules of ATP. Two molecules of ATP are produced as the net gain of glycolysis, so the grand total is 38 molecules of ATP (36 in eukaryotes). These ATP molecules may then be used in the cell for its needs.

Fermentation

Fermentation is an anaerobic process in which energy can be released from glucose even though oxygen is not available. Fermentation occurs in yeast cells, and a form of fermentation takes place in bacteria.

In yeast cells, glucose can be metabolized through cellular respiration, as in other cells. When oxygen is lacking, however, glucose is still changed to pyruvic acid via glycolysis. The pyruvic acid is first converted to acetaldehyde and then to ethyl alcohol. The net gain of ATP to the yeast cell is two molecules--the two molecules of ATP normally produced in glycolysis.

Yeasts are able to participate in fermentation because they have the necessary enzyme to convert pyruvic acid to ethyl alcohol. This process is essential because it removes electrons and hydrogen ions from NADH during glycolysis. The effect is to free the NAD so that it can participate in future reactions of glycolysis. Yeasts are therefore used in both bread and alcohol production. Alcohol fermentation is the process that yields beer, wine, and other spirits. The CO_2 given off, supplements the carbon dioxide given off during the Krebs cycle and causes bread to rise.

Biogeochemical Cycles

Energy flows directionally through ecosystems, entering as sunlight for phototrophs or as inorganic molecules for chemoautotrophs. The six most common elements associated with organic molecules—carbon, hydrogen, nitrogen, oxygen, phosphorus, and sulfur—take a variety of chemical forms and may exist for long periods in the atmosphere, on land, in water, or beneath earth's surface. Geologic processes, such as erosion, water drainage, the movement of the continental plates, and weathering, all are involved in the cycling of elements on earth. Because geology and chemistry have major roles in the study of this process, the recycling of inorganic matter between living organisms and their nonliving environment is called a biogeochemical cycle.

Carbon Cycle

Carbon is one of the most important elements to living organisms, as shown by its abundance and presence in all organic molecules. The carbon cycle exemplifies the connection between organisms in various ecosystems. Carbon is exchanged between heterotrophs and autotrophs within and between ecosystems primarily by way of atmospheric CO_2, a fully oxidized version of carbon that serves as the basic building block that autotrophs use to build multi-carbon, high-energy organic molecules such as glucose. Photoautotrophs and chemoautotrophs harness energy from the sun and from inorganic chemical compounds, respectively, to covalently bond carbon atoms together into reduced organic compounds whose energy can be later accessed through the processes of respiration and fermentation.

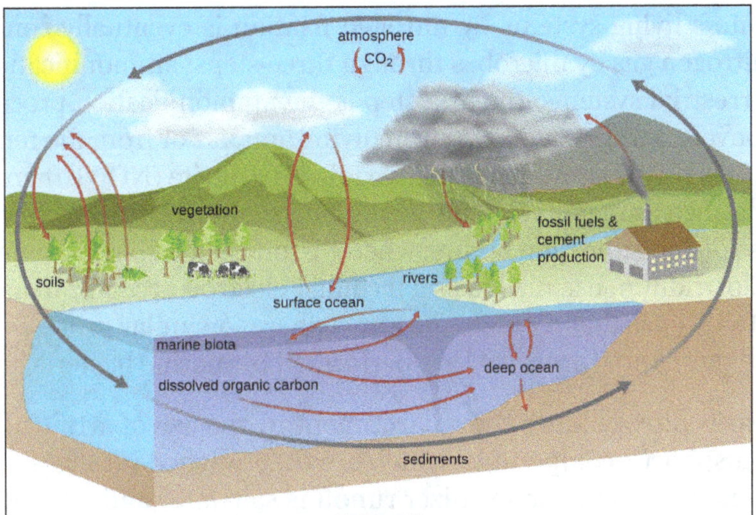

Figure: The carbon cycle- Eukaryotes participate in aerobic respiration, fermentation, and oxygenic photosynthesis. Prokaryotes participate in all the steps shown.

Overall, there is a constant exchange of CO_2 between the heterotrophs (which produce CO_2 as a result of respiration or fermentation) and the autotrophs (which use the CO_2 for fixation). Autotrophs also respire or ferment, consuming the organic molecules they form; they do not fix carbon for heterotrophs, but rather use it for their own metabolic needs.

Bacteria and archaea that use methane as their carbon source are called methanotrophs. Reduced one-carbon compounds like methane accumulate in certain anaerobic environments when CO_2 is used as a terminal electron acceptor in anaerobic respiration by archaea called methanogens. Some methanogens also ferment acetate (two carbons) to produce methane and CO_2. Methane accumulation due to methanogenesis occurs in both natural anaerobic soil and aquatic environments; methane accumulation also occurs as a result of animal husbandry because methanogens are members of the normal microbiota of ruminants. Environmental methane accumulation due to methanogenesis is of consequence because it is a strong greenhouse gas, and methanotrophs help to reduce atmospheric methane levels.

Nitrogen Cycle

Many biological macromolecules, including proteins and nucleic acids, contain nitrogen; however, getting nitrogen into living organisms is difficult. Prokaryotes play essential roles in the nitrogen cycle, transforming nitrogen between various forms for their own needs, benefiting other organisms indirectly. Plants and phytoplankton cannot incorporate nitrogen from the atmosphere (where it exists as tightly bonded, triple covalent N_2), even though this molecule composes approximately 78% of the atmosphere. Nitrogen enters the living world through free-living and symbiotic bacteria, which incorporate nitrogen into their macromolecules through specialized biochemical pathways called nitrogen fixation. Cyanobacteria in aquatic ecosystems fix inorganic nitrogen (from nitrogen gas) into ammonia (NH_3) that can be easily incorporated into biological macromolecules. Rhizobium bacteria also fix nitrogen and live symbiotically in the root nodules of legumes (such as beans, peanuts, and peas), providing them with needed organic nitrogen while receiving fixed carbon as sugar in exchange. Free-living bacteria, such as members of the genus Azotobacter, are also able to fix nitrogen.

The nitrogen that enters living systems by nitrogen fixation is eventually converted from organic nitrogen back into nitrogen gas by microbes through three steps: ammonification, nitrification, and denitrification. In terrestrial systems, the first step is the ammonification process, in which certain bacteria and fungi convert nitrogenous waste from living animals or from the remains of dead organisms into ammonia (NH_3). This ammonia is then oxidized to nitrite (NO_2^-), then to nitrate (NO_3^-), by nitrifying soil bacteria such as members of the genus Nitrosomonas, through the process of nitrification. Last, the process of denitrification occurs, whereby soil bacteria, such as members of the genera Pseudomonas and Clostridium, use nitrate as a terminal electron acceptor in anaerobic respiration, converting it into nitrogen gas that reenters the atmosphere. A similar process occurs in the marine nitrogen cycle, where these three processes are performed by marine bacteria and archaea.

Human activity releases nitrogen into the environment by the use of artificial fertilizers that contain nitrogen and phosphorus compounds, which are then washed into lakes, rivers, and streams by surface runoff. A major effect from fertilizer runoff is saltwater and freshwater eutrophication, in which nutrient runoff causes the overgrowth and subsequent death of aquatic algae, making water sources anaerobic and inhospitable for the survival of aquatic organisms.

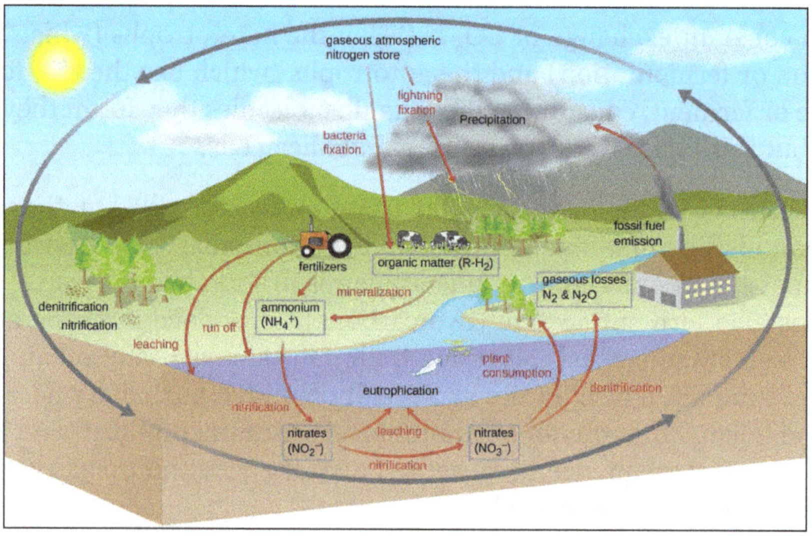

Figure: The Nitrogen cycle- Specific groups of prokaryotes each participate in every step in the cycle.

Sulfur Cycle

Sulfur is an essential element for the macromolecules of living organisms. As part of the amino acids cysteine and methionine, it is involved in the formation of proteins. It is also found in several vitamins necessary for the synthesis of important biological molecules like coenzyme A. Several groups of microbes are responsible for carrying out processes involved in the sulfur cycle. Anoxygenic photosynthetic bacteria as well as chemoautotrophic archaea and bacteria use hydrogen sulfide as an electron donor, oxidizing it first to elemental sulfur (S^o), then to sulfate (SO_2^{-4})(SO_4^{2-}). This leads to stratification of hydrogen sulfide in soil, with levels increasing at deeper, more anaerobic depths.

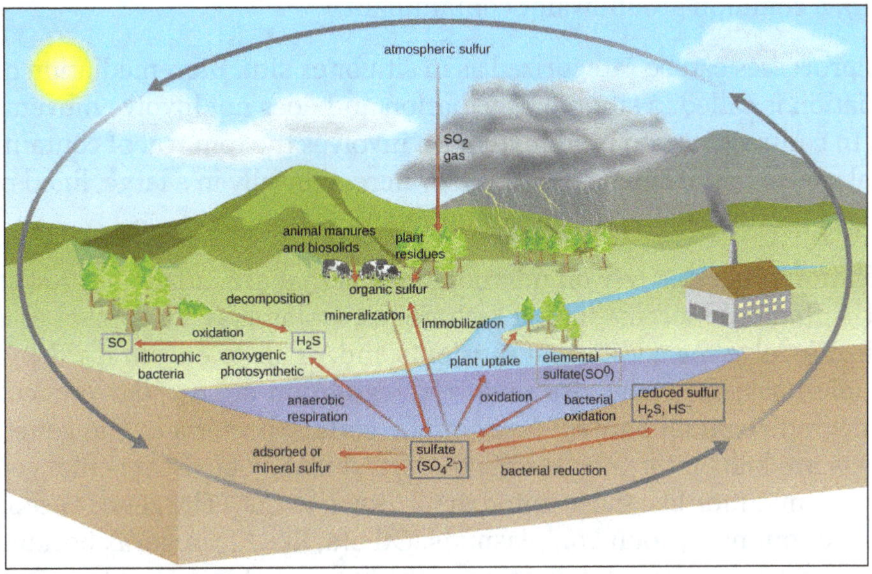

Figure: The sulfur cycle- Specific groups of prokaryotes each may participate in every step in the cycle.

Many bacteria and plants can use sulfate as a sulfur source. Decomposition dead organisms by fungi and bacteria remove sulfur groups from amino acids, producing hydrogen sulfide, returning inorganic sulfur to the environment.

Other Biogeochemical Cycles

Beyond their involvement in the carbon, nitrogen, and sulfur cycles, prokaryotes are involved in other biogeochemical cycles as well. Like the carbon, nitrogen, and sulfur cycles, several of these additional biogeochemical cycles, such as the iron (Fe), manganese (Mn), and chromium (Cr) cycles, also involve redox chemistry, with prokaryotes playing roles in both oxidation and reduction. Several other elements undergo chemical cycles that do not involve redox chemistry. Examples of these are phosphorus (P), calcium (Ca), and silica (Si) cycles. The cycling of these elements is particularly important in oceans because large quantities of these elements are incorporated into the exoskeletons of marine organisms. These biogeochemical cycles do not involve redox chemistry but instead involve fluctuations in the solubility of compounds containing calcium, phosphorous, and silica. The overgrowth of naturally occurring microbial communities is typically limited by the availability of nitrogen, phosphorus, and iron. Human activities introducing excessive amounts of iron, nitrogen, or phosphorus (typically from detergents) may lead to eutrophication.

Bioremediation

Microbial bioremediation leverages microbial metabolism to remove xenobiotics or other pollutants. Xenobiotics are compounds synthesized by humans and introduced into the environment in much higher concentrations than would naturally occur. Such environmental contamination may involve adhesives, dyes, flame retardants, lubricants, oil and petroleum products, organic solvents, pesticides, and products of the combustion of gasoline and oil. Many xenobiotics resist breakdown, and some accumulate in the food chain after being consumed or absorbed by fish and wildlife, which, in turn, may be eaten by humans. Of particular concern are contaminants like polycyclic aromatic hydrocarbon (PAH), a carcinogenic xenobiotic found in crude oil, and trichloroethylene (TCE), a common groundwater contaminant.

Bioremediation processes can be categorized as in situ or ex situ. Bioremediation conducted at the site of contamination is called in situ bioremediation and does not involve movement of contaminated material. In contrast, ex situ bioremediation involves the removal of contaminated material from the original site so that it can be treated elsewhere, typically in a large, lined pit where conditions are optimized for degradation of the contaminant.

Some bioremediation processes rely on microorganisms that are indigenous to the contaminated site or material. Enhanced bioremediation techniques, which may be applied to either in situ or ex situ processing, involve the addition of nutrients and/or air to encourage the growth of pollution-degrading microbes; they may also involve the addition of non-native microbes known for their ability to degrade contaminants. For example, certain bacteria of the genera Rhodococcus and Pseudomonas are known for their ability to degrade many environmental contaminants, including aromatic compounds like those found in oil, down to CO_2. The genes encoding their degradatory enzymes are commonly found on plasmids. Others, like Alcanivorax borkumensis, produce surfactants that are useful in the solubilization of the hydrophobic molecules found in oil, making them more accessible to other microbes for degradation.

Microbial Catabolism

Catabolism is the breakdown of complex molecules. Catabolism is the breakdown of complex substances to their constituent parts (glucose, amino acids and fatty acids) which form substrates for metabolic pathways.

Metabolism comprises two major parts; one is anabolism and the other catabolism. The Catabolism process is the action of the set of metabolic pathways that breaks down molecules into smaller units that are either oxidized to release energy or used in other anabolic reactions.

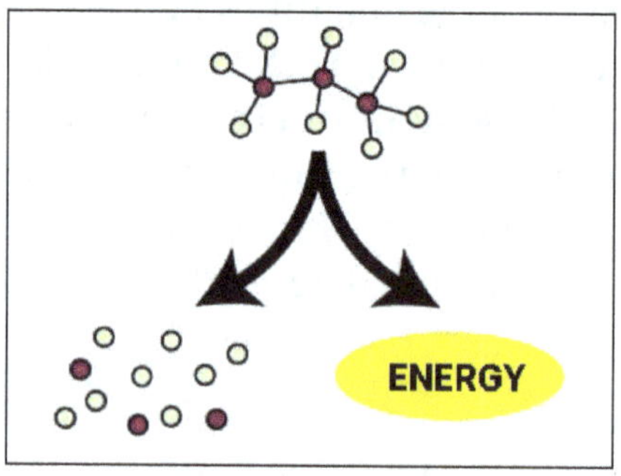

Figure: Catabolism.

The process involves the breakdown of large molecules such as polysaccharides, lipids, nucleic acids and proteins into smaller units like monosaccharides, fatty acids, nucleotides, and amino acids, respectively.

Catabolism in Different Organisms

The exact nature of catabolic reactions differs from organism to organism and can be classified based on their sources of energy and carbon which are as given below:

- Organotrophs: Use organic sources as a source of energy.
- Lithotrophs: Use inorganic substrates.
- Phototrophs: Uses sunlight as chemical energy.

Stages of Catabolism

Catabolism can be broken down into 3 main stages. The three stages are as explained as follows:

Stage 1 – Stage of Digestion

The large organic molecules of organic chemistry like proteins, lipids, and polysaccharides are digested into their smaller components outside cells. This stage acts on starch, cellulose or proteins that cannot be directly absorbed by the cells.

Stage 2 – Release of Energy

Once the molecules are broken down, these molecules are taken up by cells and converted to yet smaller molecules, usually acetyl coenzyme A, which releases some energy.

Stage 3 – Energy Stored

The released energy is stored by reducing the coenzyme nicotinamide adenine dinucleotide into NADH.

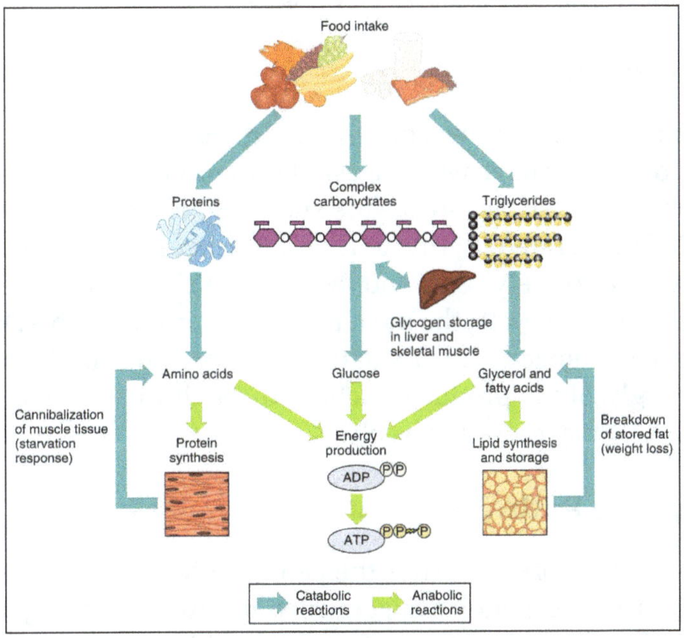

Figure: Catabolic reactions.

This process provides the chemical energy necessary for the maintenance and growth of cells. Some examples of the catabolic processes include the glycolysis, the citric acid cycle, the breakdown of muscle protein in order to use the amino acids as substrates for gluconeogenesis, the breakdown of fat in adipose tissue to fatty acids, and oxidative deamination of neurotransmitters by monoamine oxidase.

Types of Catabolism

Macromolecules such as starch, cellulose or proteins cannot be rapidly taken up by cells and must be broken into their smaller units before they can be used in cell metabolism. Several common classes of enzymes digest these polymers. These digestive enzymes include proteases that digest proteins into amino acids, as well as glycoside hydrolases that digest polysaccharides into monosaccharides. Microbes secrete digestive enzymes into their surroundings, while animals only secrete these enzymes from specialized cells in their guts. The amino acids or sugars released by these extracellular enzymes are then pumped into cells by specific active transport proteins. A simplified schematic of the catabolism of carbohydrates, proteins and fats is shown in figure below.

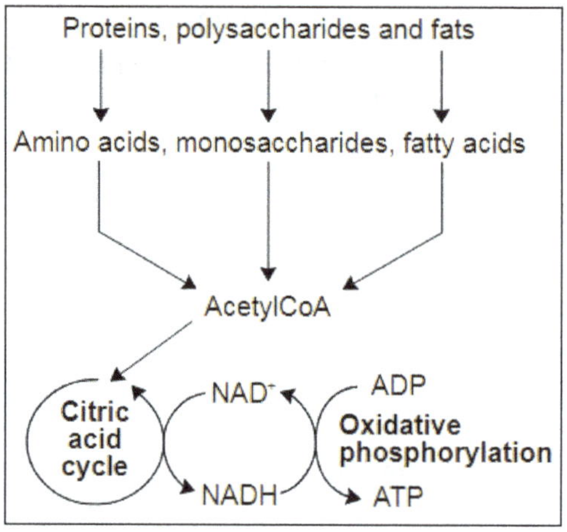

Figure: Catabolism: A simplified outline of the catabolism of proteins, carbohydrates and fats.

Carbohydrate Catabolism

Carbohydrate catabolism is the breakdown of carbohydrates into smaller units. Carbohydrates are usually taken into cells once they have been digested into monosaccharides. Once inside, the major route of breakdown is glycolysis, where sugars such as glucose and fructose are converted into pyruvate and some ATP is generated. Pyruvate is an intermediate in several metabolic pathways, but the majority is converted to acetyl-CoA and fed into the citric acid cycle. Although some more ATP is generated in the citric acid cycle, the most important product is NADH, which is made from NAD+ as the acetyl-CoA is oxidized. This oxidation releases carbon dioxide as a waste product. In anaerobic conditions, glycolysis produces lactate, through the enzyme lactate dehydrogenase re-oxidizing NADH to NAD^+ for re-use in glycolysis.

The Pentose Phosphate Pathway

An alternative route for glucose breakdown is the pentose phosphate pathway, which reduces the coenzyme NADPH and produces pentose sugars such as ribose, the sugar component of nucleic acids. Fats are catabolized by hydrolysis to free fatty acids and glycerol. The glycerol initiates glycolysis and the fatty acids are broken down by beta oxidation to release acetyl-CoA, which then is fed into the citric acid cycle. Fatty acids release more energy upon oxidation than carbohydrates because carbohydrates contain more oxygen in their structures.

Amino acids are either used to synthesize proteins and other biomolecules, or oxidized to urea and carbon dioxide as a source of energy. The oxidation pathway starts with the removal of the amino group by a transaminase. The amino group is fed into the urea cycle, leaving a deaminated carbon skeleton in the form of a keto acid. Several of these keto acids are intermediates in the citric acid cycle, for example the deamination of glutamate forms α-ketoglutarate. The glucogenic amino acids can also be converted into glucose, through gluconeogenesis.

Catabolism of Lipids and Proteins

Lipid Catabolism

Triglycerides are a form of long-term energy storage in animals. They are made of glycerol and three fatty acids. Phospholipids compose the cell and organelle membranes of all organisms except the archaea. Phospholipid structure is similar to triglycerides except that one of the fatty acids is replaced by a phosphorylated head group. Triglycerides and phospholipids are broken down first by releasing fatty acid chains (and/or the phosphorylated head group, in the case of phospholipids) from the three-carbon glycerol backbone. The reactions breaking down triglycerides are catalyzed by lipases and those involving phospholipids are catalyzed by phospholipases. These enzymes contribute to the virulence of certain microbes, such as the bacterium Staphylococcus aureus and the fungus Cryptococcus neoformans. These microbes use phospholipases to destroy lipids and phospholipids in host cells and then use the catabolic products for energy.

The resulting products of lipid catabolism, glycerol and fatty acids, can be further degraded. Glycerol can be phosphorylated to glycerol-3-phosphate and easily converted to glyceraldehyde 3-phosphate, which continues through glycolysis. The released fatty acids are catabolized in a process called β-oxidation, which sequentially removes two-carbon acetyl groups from the ends of fatty acid chains, reducing NAD^+ and FAD to produce NADH and $FADH_2$, respectively, whose electrons can be used to make ATP by oxidative phosphorylation. The acetyl groups produced during β-oxidation are carried by coenzyme A to the Krebs cycle, and their movement through this cycle results in their degradation to CO_2, producing ATP by substrate-level phosphorylation and additional NADH and $FADH_2$ molecules.

Other types of lipids can also be degraded by certain microbes. For example, the ability of certain pathogens, like Mycobacterium tuberculosis, to degrade cholesterol contributes to their virulence. The side chains of cholesterol can be easily removed enzymatically, but degradation of the remaining fused rings is more problematic. The four fused rings are sequentially broken in a multistep process facilitated by specific enzymes, and the resulting products, including pyruvate, can be further catabolized in the Krebs cycle.

Protein Catabolism

Proteins are degraded through the concerted action of a variety of microbial protease enzymes. Extracellular proteases cut proteins internally at specific amino acid sequences, breaking them down into smaller peptides that can then be taken up by cells. Some clinically important pathogens can be identified by their ability to produce a specific type of extracellular protease. For example, the production of the extracellular protease gelatinase by members of the genera Proteus and Serratia can be used to distinguish them from other gram-negative enteric bacteria. Following inoculation and growth of microbes in gelatin broth, degradation of the gelatin protein due to gelatinase production prevents solidification of gelatin when refrigerated. Other pathogens can be distinguished by their ability to degrade casein, the main protein found in milk. When grown on skim milk agar, production of the extracellular protease caseinase causes degradation of casein, which appears as a zone of clearing around the microbial growth. Caseinase production by the opportunist pathogen Pseudomonas aeruginosa can be used to distinguish it from other related gram-negative bacteria.

After extracellular protease degradation and uptake of peptides in the cell, the peptides can then be broken down further into individual amino acids by additional intracellular proteases, and each amino acid can be enzymatically deaminated to remove the amino group. The remaining molecules can then enter the transition reaction or the Krebs cycle.

Microbial Anabolism

Anabolism is the process by which the body utilizes the energy released by catabolism to synthesize complex molecules. These complex molecules are then utilized to form cellular structures that are formed from small and simple precursors that act as building blocks.

Stages of Anabolism

There are three basic stages of anabolism:

- Stage 1 involves production of precursors such as amino acids, monosaccharides, isoprenoids and nucleotides.
- Stage 2 involves activation of these precursors into reactive forms using energy from ATP.
- Stage 3 involves the assembly of these precursors into complex molecules such as proteins, polysaccharides, lipids and nucleic acids.

Sources of Energy for Anabolic Processes

Different species of organisms depend on different sources of energy. Autotrophs such as plants can construct the complex organic molecules in cells such as polysaccharides and proteins from simple molecules like carbon dioxide and water using sunlight as energy.

Heterotrophs, on the other hand, require a source of more complex substances, such as monosaccharides and amino acids, to produce these complex molecules. Photoautotrophs and photoheterotrophs obtain energy from light while chemoautotrophs and chemoheterotrophs obtain energy from inorganic oxidation reactions.

Anabolism of Carbohydrates

In these steps simple organic acids can be converted into monosaccharides such as glucose and then used to assemble polysaccharides such as starch. Glucose is made from pyruvate, lactate, glycerol, glycerate 3-phosphate and amino acids and the process is called gluconeogenesis. Gluconeogenesis converts pyruvate to glucose-6-phosphate through a series of intermediates, many of which are shared with glycolysis.

Usually fatty acids stored as adipose tissues cannot be converted to glucose through gluconeogenesis as these organisms cannot convert acetyl-CoA into pyruvate. This is the reason why when there is long term starvation, humans and other animals need to produce ketone bodies from fatty acids to replace glucose in tissues such as the brain that cannot metabolize fatty acids.

Plants and bacteria can convert fatty acids into glucose and they utilize the glyoxylate cycle, which bypasses the decarboxylation step in the citric acid cycle and allows the transformation of acetyl-CoA to oxaloacetate. From this glucose is formed.

Glycans and polysaccharides are complexes of simple sugars. These additions are made possible by glycosyltransferase from a reactive sugar-phosphate donor, such as uridine diphosphate glucose (UDP-glucose), to an acceptor hydroxyl group on the growing polysaccharide. The hydroxyl groups on the ring of the substrate can be acceptors and thus polysaccharides produced can have straight or branched structures. These polysaccharides so formed may be transferred to lipids and proteins by enzymes called oligosaccharyltransferases.

Anabolism of Proteins

Proteins are formed of amino acids. Most organisms can synthesize some of the 20 common amino acids. Most bacteria and plants can synthesize all twenty, but mammals can synthesize only the ten nonessential amino acids.

The amino acids are joined together in a chain by peptide bonds to form polypeptide chains. Each different protein has a unique sequence of amino acid residues: this is its primary structure. The polypeptide chain undergoes modifications, folding and structural changes to form the final protein.

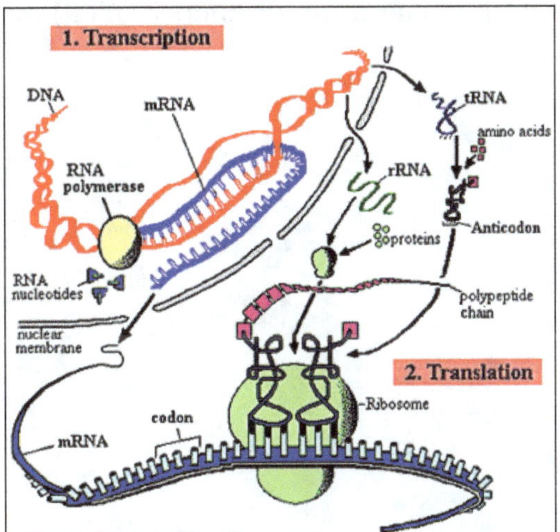

Protein synthesis

Nucleotides are made from amino acids, carbon dioxide and formic acid in pathways that require large amounts of metabolic energy.

Purines are synthesized as nucleosides (bases attached to ribose). Adenine and guanine for example are made from the precursor nucleoside inosine monophosphate, which is synthesized using atoms from the amino acids glycine, glutamine, and aspartic acid, as well as formate transferred from the coenzyme tetrahydrofolate.

Pyrimidines, like thymine and cytosine, are synthesized from the base orotate, which is formed from glutamine and aspartate.

Anabolism of Fatty Acids

Fatty acids are synthesized using fatty acid synthases that polymerize and then reduce acetyl-CoA units. These fatty acids contain acyl chains that are extended by a cycle of reactions that add the actyl group, reduce it to an alcohol, dehydrate it to an alkene group and then reduce it again to an alkane group.

In animals and fungi, all these fatty acid synthase reactions are carried out by a single multifunctional type I protein. In plants, plasmids and bacteria separate type II enzymes perform each step in the pathway.

Other lipids like terpenes and isoprenoids include the carotenoids and form the largest class of plant natural products. These compounds are made by the assembly and modification of isoprene units donated from the reactive precursor isopentenyl pyrophosphate and dimethylallyl pyrophosphate. In animals and archaea, the mevalonate pathway produces these compounds from acetyl-CoA.

Chemolithotrophy

Chemolithotrophy is a type of metabolism where energy is obtained from the oxidation of inorganic compounds. Most chemolithotrophic organisms are also autotrophic. There are two major objectives to chemolithotrophy: the generation of energy (ATP) and the generation of reducing power (NADH).

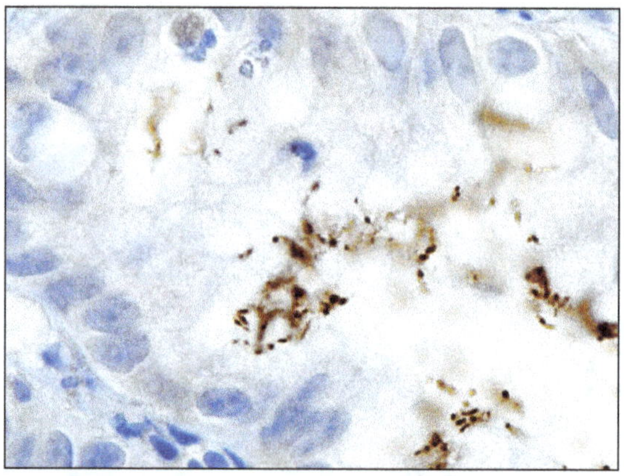

Figure: Immunohistochemical staining of H. pylori from a gastric biopsy.

Hydrogen oxidizing bacteria, or sometimes Knallgas-bacteria, are bacteria that oxidize hydrogen. These bacteria include Hydrogenobacter thermophilus, Hydrogenovibrio marinus, and Helicobacter pylori. There are both Gram positive and Gram negative knallgas bacteria.

Most grow best under microaerophilic conditions. They do this because the hydrogenase enzyme used in hydrogen oxidation is inhibited by the presence of oxygen, but oxygen is still needed as a terminal electron acceptor.

Many organisms are capable of using hydrogen (H_2) as a source of energy. While there are several mechanisms of anaerobic hydrogen oxidation (e.g. sulfate reducing- and acetogenic bacteria), hydrogen can also be used as an energy source aerobically. In these organisms, hydrogen is oxidized by a membrane-bound hydrogenase causing proton pumping via electron transfer to various quinones and cytochromes. In many organisms, a second cytoplasmic hydrogenase is used to generate reducing power in the form of NADH, which is subsequently used to fix carbon dioxide via the Calvin cycle. Hydrogen-oxidizing organisms, such as Cupriavidus necator (formerly Ralstonia eutropha), often inhabit oxic-anoxic interfaces in nature to take advantage of the hydrogen produced by anaerobic fermentative organisms while still maintaining a supply of oxygen.

Helicobacter pylori (H. pylori), previously named Campylobacter pyloridis, is a Gram-negative, microaerophilic bacterium found in the stomach. It was identified in 1982 by Barry Marshall and Robin Warren. They found that it was present in patients with chronic gastritis and gastric ulcers, conditions that were not previously believed to have a microbial cause. It is also linked to the development of duodenal ulcers and stomach cancer. However, over 80 percent of individuals infected with the bacterium are asymptomatic. It has been postulated that it may play an important role in the natural stomach ecology. More than 50% of the world's population harbor H. pylori in their upper gastrointestinal tract. Infection is more prevalent in developing countries and incidence is decreasing in Western countries. H. pylori's helix shape (from which the generic name is derived) is thought to have evolved to penetrate the mucoid lining of the stomach.

Oxidation of Reduced Sulfur Compounds

Sulfur oxidation involves the oxidation of reduced sulfur compounds, inorganic sulfur, and thiosulfate to form sulfuric acid.

Sulfur is an essential element for all life, and it is widely used in biochemical processes. In metabolic reactions, sulfur compounds serve as both fuels and respiratory (oxygen-alternative) materials for simple organisms. Sulfur is an important part of many enzymes and antioxidant molecules such as glutathione and thioredoxin.

Sulfur Oxidation

Sulfur oxidation involves the oxidation of reduced sulfur compounds such as sulfide (H_2S), inorganic sulfur (S_0), and thiosulfate ($S_2O_2^{-3}$) to form sulfuric acid (H_2SO_4). An example of a sulfur-oxidizing bacterium is Paracoccus.

Generally, the oxidation of sulfide occurs in stages, with inorganic sulfur being stored either inside or outside of the cell until needed. The two step process occurs because sulfide is a better electron donor than inorganic sulfur or thiosulfate; this allows a greater number of protons to be translocated across the membrane. Sulfur-oxidizing organisms generate reducing power for carbon dioxide fixation via the Calvin cycle using reverse electron flow—an energy-requiring process that pushes the electrons against their thermodynamic gradient to produce NADH. Biochemically, reduced sulfur compounds are converted to sulfite (SO_2^{-3}) and, subsequently, sulfate (SO_2^{-4}) by the enzyme sulfite oxidase. Some organisms, however, accomplish the same oxidation using a reversal of the APS reductase system used by sulfate-reducing bacteria. In all cases the energy liberated is transferred to the electron transport chain for ATP and NADH production. In addition to aerobic sulfur oxidation, some organisms (e.g. Thiobacillus denitrificans) use nitrate (NO^{-3}) as a terminal electron acceptor and therefore grow anaerobically.

Beggiatoa

A classic example of a sulfur-oxidizing bacterium is Beggiatoa, a microbe originally described by Sergei Winogradsky, one of the founders of environmental microbiology. Beggiatoa can be found in marine or freshwater environments. They can usually be found in habitats that have high levels of hydrogen sulfide. These environments include cold seeps, sulfur springs, sewage contaminated water, mud layers of lakes, and near deep hydrothermal vents. Beggiatoa can also be found in the rhizosphere of swamp plants. During his research in Anton de Bary's laboratory of botany in 1887, Russian botanist Winogradsky found that Beggiatoa oxidized hydrogen sulfide (H_2S) as an energy source, forming intracellular sulfur droplets. Winogradsky referred to this form of metabolism as inorgoxidation (oxidation of inorganic compounds). The finding represented the first discovery of lithotrophy.

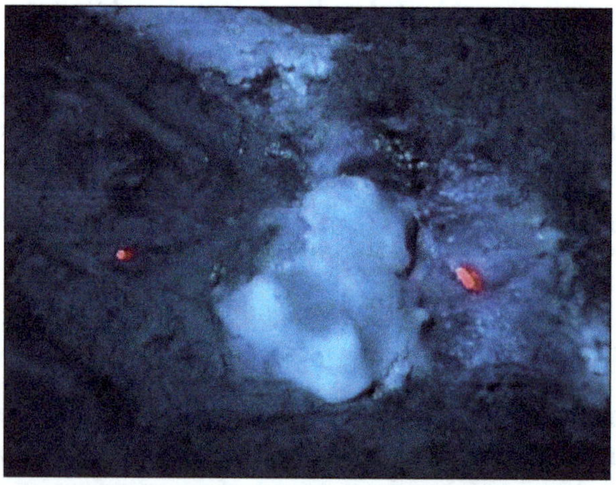

Figure: A Beggiatoa bacterial mat at the Blake Ridge: Beggiatoa spp. bacterial mat at a steep on Blake Ridge, off the coast of South Carolina. The red dots are range-finding laser beams. Beggiatoa are able to detoxify hydrogen sulfide in soil.

Beggiatoa can grow chemoorgano-heterotrophically by oxidizing organic compounds to carbon dioxide in the presence of oxygen, though high concentrations of oxygen can be a limiting factor. Organic compounds are also the carbon source for biosynthesis. Some species may oxidize hydrogen sulfide to elemental sulfur as a supplemental source of energy (facultatively litho-heterotroph). This sulfur is stored intracellularly. Some species have the ability of chemolithoautotrophic growth, using sulfide oxidation for energy and carbon dioxide as a source of carbon for biosynthesis. In this metabolic process, internal stored nitrate is the electron acceptor and reduced to ammonia.

Sulfide oxidation: $2H_2S + O_2 \rightarrow 2S + 2H_2O$

Marine autotrophic Beggiatoa species are able to oxidize intracellular sulfur to sulfate. The reduction of elemental sulfur frequently occurs when oxygen is lacking. Sulfur is reduced to sulfide at the cost of stored carbon or by added hydrogen gas. This may be a survival strategy to bridge periods without oxygen.

Iron Oxidation

Ferric iron is an anaerobic terminal electron acceptor, with the final enzyme a ferric iron reductase. Ferric iron (Fe^{3+}) is a widespread anaerobic terminal electron acceptor both for autotrophic and heterotrophic organisms. Electron flow in these organisms is similar to those in electron transport, ending in oxygen or nitrate, except that in ferric iron-reducing organisms the final enzyme in this system is a ferric iron reductase. Model organisms include Shewanella putrefaciens and Geobacter metallireducens. Since some ferric iron-reducing bacteria (e.g. G. metallireducens) can use toxic hydrocarbons such as toluene as a carbon source, there is significant interest in using these organisms as bioremediation agents in ferric iron-rich contaminated aquifers.

Figure: Iron Bacteria.

Common effects of excess iron in water are a reddish-brown color and stained laundry. Iron bacteria are a natural part of the environment in most parts of the world. These microorganisms combine dissolved iron or manganese with oxygen and use it to form rust-colored deposits. In the process, the bacteria produce a brown slime that builds up on well screens, pipes, and plumbing fixtures. Bacteria known to feed on iron are Thiobacillus ferrooxidans and Leptospirillum ferrooxidans.

Ferrous iron is a soluble form of iron that is stable at extremely low pH or under anaerobic conditions. Under aerobic, moderate pH conditions ferrous iron is oxidized spontaneously to the ferric (Fe^{3+}) form and is hydrolyzed abiotically to insoluble ferric hydroxide ($Fe(OH)_3$). There are three distinct types of ferrous iron-oxidizing microbes. The first are acidophiles, such as the bacteria Acidithiobacillus ferrooxidans and Leptospirillum ferrooxidans, as well as the archaeon Ferroplasma. These microbes oxidize iron in environments that have a very low pH and are important in acid mine drainage. The second type of microbes oxidizes ferrous iron at cirum-neutral pH. These micro-organisms (for example Gallionella ferruginea or Leptothrix ochracea) live at the oxic-anoxic interfaces and are microaerophiles. The third type of iron-oxidizing microbes is anaerobic photosynthetic bacteria such as Rhodopseudomonas, which use ferrous iron to produce NADH for autotrophic carbon dioxide fixation. Biochemically, aerobic iron oxidation is a very energetically poor process which therefore requires large amounts of iron to be oxidized by the enzyme rusticyanin to facilitate the formation of proton motive force. Like sulfur oxidation, reverse electron flow must be used to form the NADH used for carbon dioxide fixation via the Calvin cycle.

Although ferric iron is the most prevalent inorganic electron acceptor, a number of organisms (including the iron-reducing bacteria) can use other inorganic ions in anaerobic respiration. While these processes may often be less significant ecologically, they are of considerable interest for bioremediation, especially when heavy metals or radionuclides are used as electron acceptors. Examples include:

- Manganic ion (Mn^{4+}) reduction to manganous ion]] (Mn^{2+}).
- Selenate (SeO_2^{-4}) reduction to selenite (SeO_2^{-3}) and selenite reduction to inorganic selenium (Se_0).
- Arsenate (AsO_3^{-4}) reduction to arsenite (AsO_3^{-3}).
- Uranyl ion ion (UO_2^{+2}) reduction to uranium dioxide (UO_2).

Nitrification

Nitrification is the process by which ammonia (NH_3) or ammonium (NH_4^+) is converted to nitrite (NO_2^-) and then nitrate (NO_3^-) by bacteria.

Process of Nitrification

Nitrification is the process by which ammonia (NH_3) or ammonium (NH4$^+$) is converted to nitrate (NO_3^-). Nitrification is the net result of two distinct processes: oxidation of ammonium to nitrite (NO_2^-) by nitrosifying or ammonia-oxidizing bacteria and oxidation of nitrite (NO_2^-) to nitrate (NO_3^-) by the nitrite-oxidizing bacteria. Nitrification is an important step in the nitrogen cycle in soil. Nitrification is an aerobic process performed by small groups of autotrophic bacteria and archaea.

Chemistry of Nitrogen Compound Oxidation

Nitrification is a process of nitrogen compound oxidation (effectively, loss of electrons from the nitrogen atom to the oxygen atoms):

$$2\,NH_4^+ + 3\,O_2 \rightarrow 2\,NO_2^- + 2\,H_2O + 4\,H^+\,(Nitrosomonas)$$

$$2\,NO_2^- + O_2 \rightarrow 2\,NO_3^-\,(Nitrobacter,\ Nitrospina)$$

$$NH_3 + O_2 \rightarrow NO_2^- + 3H^+ + 2e^-$$

$$NO_2^- + H_2O \rightarrow NO_3^- + 2H^+ + 2e^-$$

Both of these processes are extremely energetically poor, which leads to very slow growth rates for both types of organisms.

Ammonium Oxidation

The transformation of ammonia to nitrite is usually the rate limiting step of nitrification. Biochemically, ammonium oxidation occurs by the stepwise oxidation of ammonium to hydroxylamine (NH_2OH) by the enzyme ammonium monooxygenase in the cytoplasm, followed by the oxidation of hydroxylamine to nitrite by the enzyme hydroxylamine oxidoreductase in the periplasm. Electron and proton cycling are very complex, but as a net result only one proton is translocated across the membrane per molecule of ammonium oxidized.

Nitrite Reduction

Nitrite reduction is much simpler, with nitrite being oxidized by the enzyme nitrite oxidoreductase coupled to proton translocation by a very short electron transport chain, again leading to very low growth rates for these organisms. Oxygen is required in ammonium and nitrite oxidation, meaning that both nitrosifying and nitrite-oxidizing bacteria are aerobes. As in sulfur and iron oxidation, NADH for carbon dioxide fixation using the Calvin cycle is generated by reverse electron flow, thereby placing a further metabolic burden on an already energy-poor process.

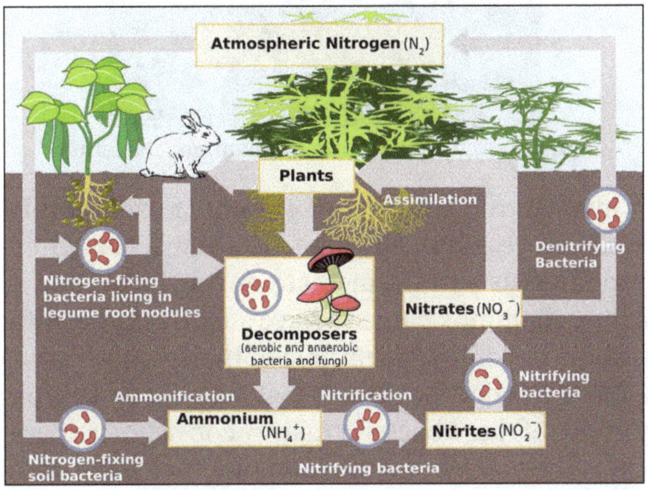

Figure: The nitrogen cycle: Schematic representation of the flow of nitrogen through the environment. The importance of bacteria in the cycle is immediately recognized as being a key element in the cycle, providing different forms of nitrogen compounds assimilable by higher organisms.

Human Applications of Nitrification

Nitrification is important in agricultural systems, where fertilizer is often applied as ammonia. Nitrification also plays an important role in the removal of nitrogen from municipal wastewater. The conventional removal is nitrification, followed by denitrification.

Anammox

Anammox, an abbreviation for ANaerobic AMMonium OXidation, is a globally significant microbial process of the nitrogen cycle.

Anammox, an abbreviation for ANaerobic AMMonium OXidation, is a globally significant microbial process of the nitrogen cycle. The bacteria mediating this process were identified in 1999, and at the time were a great surprise to the scientific community. Anammox takes place in many natural environments, contributing up to 50% of the dinitrogen gas produced in the oceans. In this biological process, nitrite and ammonium are converted directly into dinitrogen gas. The overall catabolic reaction is:

$$NH_4^+ + NO_2^- \rightarrow N_2 + 2H_2O$$

Figure: Enrichment Culture of Anammox Bacterium.

This form of metabolism involves the coupling of ammonia oxidation to nitrite reduction. Since oxygen is not required for the process, these organisms are strict anaerobes. Amazingly, hydrazine (N_2H_4 – rocket fuel) is produced as an intermediate during anammox metabolism. To deal with the high toxicity of hydrazine, anammox bacteria have a hydrazine-containing intracellular organelle called the anammoxasome (a compartment inside the cytoplasm which is the locus of anammox catabolism), which is surrounded by an unusual and highly compact ladderane lipid membrane. Further, the membranes of these bacteria mainly consist of ladderane lipids so far unique in biology. Of special interest is the conversion to hydrazine (normally used as a high-energy rocket fuel, and poisonous to most living organisms) as an intermediate. A final striking feature of the organism is the extremely slow growth rate. The doubling time is nearly two weeks. The anammox process was originally found to occur only from 20 °C to 43 °C but more recently, anammox has been observed at temperatures from 36 °C to 52 °C in hot springs and 60 °C to 85 °C at hydrothermal vents located along the Mid-Atlantic Ridge.

Anammox organisms are autotrophs although the mechanism for carbon dioxide fixation is still unclear. Because of this property, these organisms could be used industrially to remove nitrogen in wastewater treatment processes. The bacteria that perform the anammox process belong to the bacterial phylum Planctomycetes (e.g. Candidatus Brocadia anammoxidans), of which Planctomyces and Pirellula are the best known genera. Currently five genera of anammox bacteria have been (provisionally) defined: Brocadia, Kuenenia, Anammoxoglobus, Jettenia (all fresh water species), and Scalindua (marine species).

Benzoate Catabolism

Rhodobacter sphaeroides is able to produce hydrogen from a wide range of organic compounds (chiefly organic acids) and light. Benzoate catabolism is a series of chemical reactions resulting in the breakdown of benzoate. The purple non-sulphur (PNS) bacteria Rhodobacter sphaeroides is able to produce hydrogen from a wide range of organic compounds (chiefly organic acids) and light. The photo-system required for hydrogen production in Rhodobacter (PS-I) differ from its oxygenic photosystem (PS-II) due to the requirement of organic acids and the inability to oxidize water. In the absence of water-splitting, photosynthesis is anoxygenic. Therefore, hydrogen production is sustained without inhibition from generated oxygen. In PNS bacteria, hydrogen production is due to catalysis by nitrogenase. Hydrogenases are also present but the production of hydrogen by [FeFe]-hydrogenase is less than ten times the hydrogen uptake by [NiFe]-hydrogenase. Only under nitrogen-deficient conditions is nitrogenase activity sufficient to overcome uptake hydrogenase activity, resulting in net generation of hydrogen.

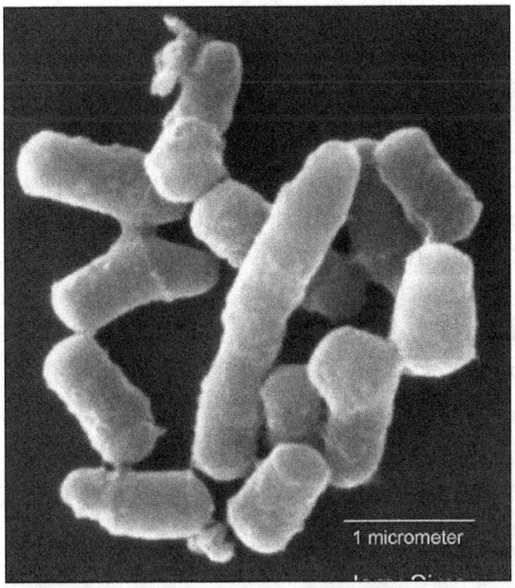

Figure: Scanning electron micrograph of Rhodococcus sp. strain Q1: Rhodococcus sp. strain Q1 grown on quinoline – the organism can use quinoline as a sole source of carbon, nitrogen and energy, tolerating concentrations up to 3.88 millimoles per liter.

Rhodococcus is a genus of aerobic, nonsporulating, non-motile Gram-positive bacteria closely related to Mycobacteria and Corynebacteria. While a few species are pathogenic, most are benign and have been found to thrive in a broad range of environments, including soil, water, and eukaryotic cells. Fully sequenced in October 2006, the genome is known to be 9.7 megabasepairs long and 67% G/C. Strains of Rhodococcus are applicably important owing to their ability to catabolize a wide range of compounds and produce bioactive steroids, acrylamide, and acrylic acid, and their involvement in fossil fuel biodesulfurization. This genetic and catabolic diversity is not only due to the large bacterial chromosome, but also to the presence of three large linear plasmids. Rhodococcus is also an experimentally advantageous system owing to a relatively fast growth rate and simple developmental cycle. However, as it stands now, Rhodococcus is not well characterized. Another important application of Rhodococcus comes from bioconversion, using biological systems to convert cheap starting material into more valuable compounds. This use of Rhodococcus is borne out of its ability to metabolize harmful environmental pollutants, such as toluene, naphthalene, herbicides, and PCBs. Rhodococci typically metabolize aromatic substrates by first oxygenating the aromatic ring to form a diol (two alcohol groups). Then, the ring is cleaved with intra/extradiol mechanisms, opening the ring and exposing the substrate to further metabolism. Since the chemistry here is very stereospecific, the diols are created with predictable chirality. While controlling the chirality of chemical reaction presents a significant challenge for synthetic chemists, biological processes can be used instead to faithfully produce chiral molecules in cases where direct chemical synthesis is infeasible or inefficient. An example of this is the use of Rhodococcus to produce indene, a precursor to the AIDS drug Crixivan, a protease inhibitor, and containing two of the five chiral centers needed in the complex.

Polycyclic Aromatic Hydrocarbons

Polycyclic aromatic hydrocarbons are potent atmospheric pollutants that consist of fused aromatic rings and do not contain heteroatoms.

PAHs

Polycyclic aromatic hydrocarbons (PAHs), also known as poly-aromatic hydrocarbons or polynuclear aromatic hydrocarbons are seen in. PAHs are potent atmospheric pollutants that consist of fused aromatic rings and do not contain heteroatoms or carry substituents. Naphthalene is the simplest example of a PAH. PAHs occur in oil, coal, and tar deposits, and are produced as byproducts of fuel burning (whether fossil fuel or biomass). As a pollutant, they are of concern because some compounds have been identified as carcinogenic, mutagenic, and teratogenic. PAHs are also found in cooked foods—studies have found PAHs in meat cooked at high temperatures such as grilling or barbecuing, and in smoked fish. They are also found in the interstellar medium, comets, and meteorites. PAHs are a candidate for the molecule acted as a basis for the earliest forms of life. In graphene the PAH motif is extended to large 2D sheets.

Figure: Polycyclic Aromatic Hydrocarbons: An image showing three examples of polycyclic aromatic hydrocarbons. Clockwise from top left, the molecules are: benz[e]acephenanthrylene, pyrene and dibenz[a,h]anthracene.

PLFA Analysis

Phospholipid -derived fatty acids (PLFA) are widely used in microbial ecology as chemotaxonomic markers of bacteria and other organisms. Phospholipids are the primary lipids composing cellular membranes. They can be esterified to many types of fatty acids. Once the phospholipids of an unknown sample are esterfied, the composition of the resulting PLFA can be compared to the PLFA of known organisms to determine the identity of the sample organism. PLFA analysis may be combined with stable isotope probing to determine which microbes are metabolically active in a sample.

The basic premise for PLFA analysis is that as individual organisms (especially bacteria and fungi) die, phospholipids are rapidly degraded and the remaining phospholipid content of the sample is assumed to be from living organisms. As the phospholipids of different groups of bacteria and fungi contain a variety of somewhat unique fatty acids, they can serve as useful biomarkers for such groups. PLFA profiles and composition can be determined by purifying the phospholipids and then cleaving the fatty acids for further analysis.

Bioremediation

Bioremediation is the use of micro-organism metabolism to remove pollutants. Technologies can be generally classified as in situ or ex situ. *In situ* bioremediation involves treating the contaminated material at the site, while ex situ involves the removal of the contaminated material to be treated elsewhere. Some examples of bioremediation related technologies are phytoremediation, bioventing, bioleaching, landfarming, bioreactor, composting, bioaugmentation, rhizofiltration, and biostimulation.

Bioremediation can occur on its own (natural attenuation or intrinsic bioremediation) or can be spurred on via the addition of fertilizers to increase the bioavailability within the medium (biostimulation). Recent advancements have found success by adding matched microbe strains to the medium to enhance the resident microbe population's ability to break down contaminants. Microorganisms used to perform the function of bioremediation are known as bioremediators. Not all contaminants, however, are easily treated by bioremediation using microorganisms. For example, heavy metals such as cadmium and lead are not readily absorbed or captured by microorganisms. The assimilation of metals such as mercury into the food chain may worsen matters.

Phytoremediation

Phytoremediation is useful in these circumstances because natural plants or transgenic plants are able to bioaccumulate these toxins in their above-ground parts, which are then harvested for removal. The heavy metals in the harvested biomass may be further concentrated by incineration or recycled for industrial use. The elimination of a wide range of pollutants and wastes from the environment requires increasing our understanding of the relative importance of different pathways and regulatory networks to carbon flux in particular environments and for particular compounds, and they will certainly accelerate the development of bioremediation technologies and biotransformation processes.

Mycoremediation

Mycoremediation is a form of bioremediation, the process of using fungi to degrade or sequester contaminants in the environment. Stimulating microbial and enzyme activity, mycelium reduces toxins *in situ*. Some fungi are hyperaccumulators, capable of absorbing and concentrating heavy metals in the mushroom fruit bodies. One of the primary roles of fungi in the ecosystem is decomposition, which is performed by the mycelium. The mycelium secretes extracellular enzymes and acids that break down lignin and cellulose, the two main building blocks of plant fiber. These organic compounds are composed of long chains of carbon and hydrogen, structurally similar to many organic pollutants. The key to mycoremediation is determining the right fungal species to target a specific pollutant.

Gallaecimonas is a recently described genus of bacteria. It is a Gram-negative, rod-shaped, halotolerant bacterium in the class Gammaproteobacteria. It can degrade high molecular mass polycyclic aromatic hydrocarbons of 4 and 5 rings. The 16S rRNA gene sequences of the type strain CEE_131 (T) proved to be distantly related to those of Rheinheimera and Serratia.

Phototrophy

Phototrophy refers to the utilization of solar electromagnetic radiation for energy conservation. In most cases, but not all phototrophy also implies a close coupling to the energy requirement for reduction of CO_2 for assimilation into organic matter according to the general scheme:

$$2H_2A + CO_2 \rightarrow 2\left[CH_2O\right] + 2A$$

Where H_2A represents an external electron donor and A is an oxidized product in a process referred to as photosynthesis.

The most important examples of phototrophy are based on chlorophylls or bacteriochlorophylls. In both cases a *reaction center* consists of a chlorophyll or bacteriochlorophyll molecule surrounded by other light-harvesting molecules, including carotenoids, and that absorb light in the 400 nm–550 nm range and sometimes other accessory photosynthetic pigments, phaeophytin and an electron transport chain. The wavelengths exploited are within the range of 400 nm–1000 nm, but different regions of this band are exploited by different chlorophylls and accessory photosynthetic pigments, so there is a differential utilization of the light spectrum among different types of phototrophs based on the pigments they possess. The reaction centers reside in the membranes of intracellular vesicles or tubular structures, which either constitute invaginations of the cell membrane (in purple bacteria and cyanobacteria) or are made up of special (non-unit) membranes (in green non-sulfur phototrophic bacteria, e.g., Chlorobium).

The principal process involves delivery of an electron by a photoactivated chlorophyll or bacteriochlorophyll to a primary acceptor (phaeophytin: a chlorophyll or bacteriochlorophyll deprived of the Mg atom) which passes the electron on to a membrane-bound electron carrier chain (including FeS proteins, ubiquinones and cytochromes) and then back to the chlorophyll molecule in a cyclic manner. This acts like a proton pump which is coupled to ATP synthesis by membrane phosphorylation (cyclic photophosphorylation). This is quite analogous to respiration, only in phototrophs the electron source is a photoactivated chlorophyll rather than NADH. For photosynthesis the electrons can also be used for the production of reduced nicotinamide dinucleotide phosphate (NADPH), which is in turn used for the reductive assimilation of CO_2.

Bacteria performing so-called anoxgenic photosynthesis depend on electron donors other than

H_2O, and thus do not generate O_2. Such anoxygenic bacteria are found within four unrelated groups of bacteria; these are characterized by, among other features, their photosynthetic pigments. The purple sulfur bacteria (phylum Proteobacteria) have bacteriochlorophyll *a* with an infrared absorption maximum at 825–890 nm or bacteriochlorophyll *b* (1020–1040 nm); different groups of purple sulfur bacteria have different combinations of carotenoids. The purple sulfur bacteria (e.g. Thiocapsa, Thiopedia, and Chromatium) typically use reduced sulfur as an electron donor for photosynthesis according to:

$$2H + +2HS + CO_2 \rightarrow [CH_2O] + 2S^o + H_2O$$

and

$$5H_2O + 2So + 3CO_2 \rightarrow 3[CH_2O] + 2SO_4^{2-} + 4H^+$$

The cells often store elemental S. These organisms occur in and often dominate sulfidic habitats exposed to light, e.g., the hypolimnion of some lakes and some microbial mats. They are autotrophs, and to a varying degree O_2-tolerant, but some are strict anaerobes and this, as well as differences in sulfide tolerance, explains the variability in their occurrence in nature. Recently, an isolate closely related to the purple sulfur bacterium, Ectothiorhodopspira, has also been found to use reduced arsenic in the form of arsenite (AsO_3^{3-}) as an electron donor for photosynthesis, producing arsenate (AsO_4^{3-}) as an end product.

The term purple non-sulfur bacteria (phylum Proteobacteria; e.g., Rhodopseudomonas, Rhodospirillum) is not accurate in that many isolates can utilize sulfide as an electron donor in photosynthesis, although they are more sensitive to higher sulfide levels than are the purple sulfur bacteria, and they do not store elemental S in their cells. They otherwise use H_2, some organic compounds and Fe^{2+} as electron donors. Most recently it has been found that at least one purple sulfur bacterium and a purple non-sulfur bacterium can use nitrite as an electron donor with NO_3^- as an end product. The purple non-sulfur bacteria can also take up low molecular weight organic compounds such as acetate; in the dark, they prefer microaerobic conditions and perform aerobic respiration. Purple non-sulfur bacteria are common in surface sediments.

Some members of the purple non-sulfur bacteria (within the α-Proteobacteria) have lost their ability to photosynthesize, but retained the capacity for cyclic photophosphorylation. Thus they have an absolute requirement for organic substrates for growth, but can conserve energy from light, which gives them a competitive edge relative to other heterotrophic bacteria, because they do not have to use as much of their organic substrates for dissimilatory metabolism. This presumably allowed for improved survival during starvation periods. Such aerobic photoheteroptrophs have been known for some time, but it has only more recently been recognized as very numerous in the marine water column, making up more than 10% of the bacterial biota. These bacteria are often easy to isolate from seawater and their names – Erythrobacter and Roseobacter – reveal that their characteristic red colour had been noticed before their phototrophic ability was understood.

A quite different type of phototrophy occurs in a variety of bacteria – both among Archaea and Bacteria. In the 1970s an opsin protein was discovered in the cell membrane of a halophile archaeum, Halobacter. Opsins constitute a group of light-sensitive proteins. When exposed to light they act as proton pumps expelling protons out of the cell creating an electrochemical gradient coupled to

ATP-generation; the Halobacter opsin was called bacteriorhodopsin. A related protein discovered in Proteobacteria has been called proteorhodopsin. Opsins (such as rhodopsin) also constitute the light-sensitive pigments of animal eyes and mediate light responses in unicellular eukaryotes. It is now known that opsins play a role in light sensing in different kinds of bacteria, and that they are also used for bacterial energy conservation via ATP synthesis. This is a widespread property among disparate groups of marine bacterioplankton. Most likely the wide distribution of opsins among different bacterial taxa reflects horizontal gene transfer, the assessment of which illustrates the power and value of genomic and metagenomic approaches for microbial biogeochemistry.

Among the bacteria only the cyanobacteria perform oxygenic photosynthesis. Oxygenic photosynthesis is the most important phototrophic process in the present biosphere, and has been that for perhaps the last three billion years of Earth's history. It caused the rise of an O_2-containing atmosphere that among other things allowed for the evolution of multi-cellular life. Oxygenic photosynthesis also generates the bulk of chemical energy in the form of organic matter that drives the rest of the biosphere. Photosynthesis of eukaryotic organisms may play a larger role than that of cyanobacteria in the extant biosphere (especially terrestrial systems), but then again all evidence shows that eukaryotic chloroplasts originated as intracellular cyanobacterial symbionts.

Cyanobacteria (and chloroplasts) use chlorophyll *a* as their principle photosynthetic pigment; accessory photosynthetic pigments include the phycobilins (phycocyanin and phycoerythrin) that allow cyanobacteria to utilize the orange and green regions of the light spectrum. Cyanobacteria are important in shallow water sediments and dominate "cyanobacterial mats" – especially the filamentous forms, such as Oscillatoria, Lyngbya, Microcoleus, and Spirulina; however, they also occur in mats and sediments in unicellular forms. Cyanobacteria are also important as primary producers in the plankton; in this case it is mainly the unicellular types (Synechococcus) that dominate. However, mass occurrences of colonial forms such as Trichodesmium and Nodularia may also occur frequently. In addition, cyanobacteria colonize terrestrial habitats including soils and cracks in rocks (as endoliths) even in very harsh environments. Some cyanobacteria live symbiotically with eukaryotes, for example they are the photosynthetic component of some types of lichens.

Sediment dwelling cyanobacteria are frequently exposed to sulfide, which inhibits photosystem II thus preventing O_2- generation. Photosystem I, however, remains intact and under such circumstances some forms may use sulfide as an electron donor in their photosynthesis.

The cyanobacterium Acaryochloris is unusual in possessing chlorophyll *d* that allows it to utilize near infrared light for oxygenic photosynthesis. It has been found beneath crustose corralline algae in coral reefs.

The prochlorophytes are small unicellular forms; they also contain chlorophyll *b* (like chloroplasts of green algae and plants). Originally discovered as symbionts in the tunic of certain species of tunicates, they have more recently proven to be very important in marine plankton – in particular in oceanic sites in the deeper part of the photic zone. Phototrophic metabolism has evolved twice, via chlorophototrophy and retinalophototrophy.

Chlorophototrophy

Named for the chlorophyll and bacteriochlorophyll pigments that absorb light, chlorophototrophs drive both energy metabolism and redox chemistry via light. Found in cyanobacteria and at least seven other phototrophic clades of bacteria it is responsible for the vast majority of primary production of biomass on Earth as well as much of the energy metabolism of organisms which possess it. Approximately, teramoles of carbon are fixed by chlorophototrophs annually, primarily via oxygenic photosynthesis. The functional unit of the chlorophototrophic machinery is the photochemical reaction center or RC. These large membrane-bound protein complexes are all descended from an ancestral homodimer, with some diversifying into heterodimers and some accumulating numerous accessory subunits,. All chlorophototrophic reaction centers push electrons to more reducing potentials via chlorophyll and bacteriochlorophyll photochemistry, either passing these electrons to electron carriers which can be used to fix biomass or energizing an electron transport chain to produce biologically available energy. Electrons may be pulled from elsewhere in metabolism via soluble cytochromes, or in the case of cyanobacterial Photosystem II, water itself. Chlorophyll and bacteriochlorophyll pigments are biochemically derived from porphyrins and evolutionarily related to heme, as indicated by the similarity of their biosynthesis. Three central pairs of chlorophyll molecules in a transmembrane protein core represent the conserved engine of charge separation with one photo-excited chlorophyll donating an electron to another. Additional 'antenna' chlorophylls in each reaction center allow absorption of light with a higher cross-section per reaction center, with energy transferred from chlorophyll to chlorophyll via resonance transfer. The mass of the conserved core reaction center is approximately kilodaltons and when including these integrated antennas it can reach more than kilodaltons. Light-gathering capacity is further enhanced by the presence a remarkably diverse array of independently-evolved pigment-bearing accessory antenna complexes which further transfer their absorbed energy into the reaction center.

Chlorophototrophy is found only in eubacteria and in eukaryotes that have taken up photosynthetic cyanobacteria as plastid organelles, with no known archaeal chlorophototrophs. The distribution of chlorophototrophy within the eubacteria is patchy with chlorophototrophic clades scattered across the bacterial tree. Horizontal gene transfer is likely responsible for at least some of the distribution of chlorophototrophy across the tree of life with transfer positively identified into the Gemmatimonadetes, and within clades of the Proteobacteria and Chloroflexi. However this process is rare at best with horizontal transfer requiring over genes to move between species, and the relative importance of horizontal versus vertical transfer outside these examples is ambiguous.

The chlorophototrophic machinery has diversified significantly over time, with different lineages containing machinery that while operating from the same mechanistic basis has been adapted for different purposes. The deepest split in the evolutionary tree of photochemical reaction center proteins is that between type I and type II reaction centers. Type I reaction centers contain iron-sulfur clusters and are tuned to more reducing redox potentials, pushing electrons from cytochromes or other soluble electron carriers to ferredoxin using light energy. Type II reaction centers are tuned to more oxidizing redox potentials, boosting electrons to membrane soluble quinones from cytochromes or, in the case of cyanobacterial photosystem II, directly from water. While type I reaction centers produce a highly reduced electron carrier capable of driving either carbon fixation pathways or energy metabolism, the quinone reduced by type II reaction centers cannot drive carbon fixation directly and instead can only directly drive an electron transport chain, typically consisting of a cytochrome bc complex.

Retinalophototrophy

Retinalophototrophy, the second independent origin of phototrophy, was only discovered in the 1970s via investigation of the phototrophic mechanism of haloarchaea. The retinalophototrophic system is far simpler than chlorophototrophy, consisting of a single 26-28 kilodalton trans membrane protein, known as a microbial or 'type-' rhodopsin. It is covalently bound to a single pigment molecule known as retinal, derived from the oxidative splitting of a carotenoid via a dioxygenase. In a few cases, such as the xanthorhodopsins, a single additional carotenoid molecule is bound to the exterior of the protein and functions as a miniature integral 'antenna'.

Microbial rhodopsins directly pump protons across a cell membrane rather than engaging in redox chemistry. Light-driven isomerization of the retinal pigment pumps a single proton per absorbed photon across the membrane through the rhodopsin channel, meaning the system is self-contained and does not require additional electron transport chain components to extract energy.

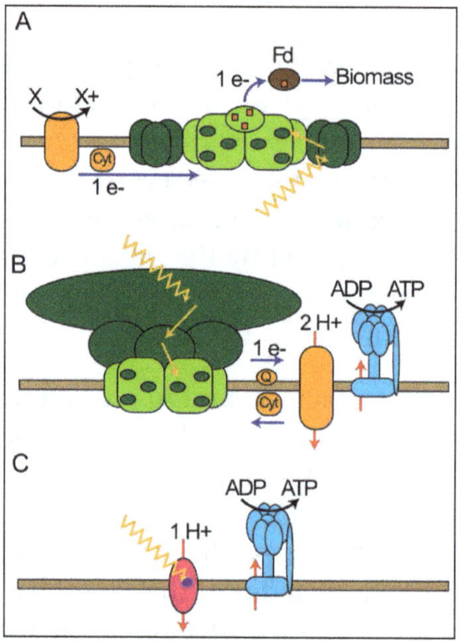

Figure: Simplified illustration of the three main types of phototrophic metabolism.

A) Chlorophototrophy, type I reaction center. A photon is absorbed by one of a diverse array of antenna complexes (dark green) and passed as an exciton via Forster resonance to chlorophyll or bacteriochlorophyll molecules (small dark green spots) within the dimeric photosynthetic reaction center (light green). A type I reaction center is illustrated acquiring an electron via a cytochrome derived from an environmental reducing agent, boosting it via light energy to a low redox potential, and passing it via iron-sulfur clusters (red) to ferredoxin (brown) which can be used to build biomass via carbon and nitrogen fixation.

B) Chlorophototrophy, type II reaction center illustrated passing electrons to a quinone electron acceptor, allowing for simple cyclic electron transfer via cytochrome bc (complex III) (orange) and the pumping of two protons per absorbed photon. C) Retinalophototrophy. A single molecule of retinal (purple) is bound to a microbial rhodopsin membrane protein (pink). Absorption of a photon causes one proton to be pumped the exterior of the cell, upon which it can participate in chemiosmotic ATP production via the membrane ATP synthase (blue).

Some rhodopsins, not directly involved in phototrophy, are also capable of pumping ions such as chloride or sodium and others function as light sensors. While there are no known autotrophs able to fix biomass from CO using only the energy derived from microbial rhodopsins, the energy generated by this system appears to be quite important for many photoheterotrophs. This energy can prevent starvation in marine bacteria, and is extensively used to supplement heterotrophic metabolism: the quantity of light absorbed by retinalophototrophs in the ocean is thought to be at least as large as that absorbed by chlorophototrophs.

The phylogenetic ubiquity of microbial rhodopsins, in contrast to the patchy distribution of chlorophototrophy, has only been fully appreciated in the last two decades. Approximately half of marine bacterial cells, from many taxa, bear diverse bacterial rhodopsin genes. In addition to haloarchaea, they are present in marine bacteria, marine archaea, fungi, and heterotrophic marine eukaryotes. They are known to acidify cellular compartments via pumping protons, and in some taxa are among the most highly expressed proteins, contributing significantly to the cell's energy budget. Rhodopsins have even been discovered in metagenomes of Heimdallarchaea, a member of the Asgard archaea considered a likely sister to the archaeal ancestor of eukaryotes and in numerous marine viruses.

Microbial rhodopsins are exemplars of horizontal gene transfer, explaining its cosmopolitan distribution across the tree of life. If a microbe contains a functional carotenoid synthesis pathway, retinalophototrophy may be transferred into the cell via a simple two-gene cassette consisting of the rhodopsin itself and an enzyme that oxidatively cleaves a carotenoid into retinal. If no carotenoid synthesis pathway exists, a total of five genes are required, constituting a basic carotenoid synthesis pathway alongside these genes. Gene cassettes of these types are widely observed in bacteria and archaea. Due to this ease of horizontal gene transfer, the evolutionary origin of microbial rhodopsins remains unclear.

Mechanistic Comparison of Phototrophic Systems

The differences between chlorophototrophy and retinalophototrophy are manifold. They represent independent origins of phototrophic metabolism, derived from different metabolic cofactors shaped into photoactive pigments and representing different trade-offs and strategies in the space of possible phototrophic metabolisms. By abstracting away from their fine details and looking at gross compositions and the products of their metabolisms the major differences between them may be understood more easily.

Table: Attributes of Chlorophototrophy and Retinalophototrophy.

	Chlorophototrophy	Microbial Rhodopsins
Distribution	Bacteria (and plastids)	Bacteria, Archaea, Eukarya
Active unit	≤350 kDa dimeric reaction center	single ~27 kDa protein
Required unit	~30	2 to 5
Mechanism	Electron transport chain	Direct proton pump
Pigment	Chlorophyll/ Bacteriochlorophyll	Retinal
Pigments evolved via	Porphyrins	Carotenoids

Antenna pigments	Diverse and abundant	One carotenoid in xanthorhodopsins
Products per cycle	One electron (or ~2 protons pumped)	One proton pumped
Used for	Energy, Carbon and Nitrogen fixation	Energy
Speed	Up to 350 electrons s^{-1} RC^{-1}	Up to 50 protons s^{-1}

Comparison of ecologically and evolutionarily relevant differences between chlorophototrophic reaction-center-based and retinalophototrophic microbial-rhodopsin-based phototrophy.

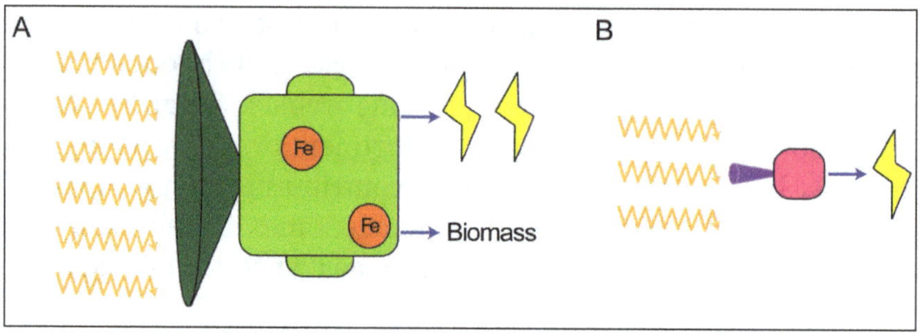

Figure: Schematic illustrating relevant functional differences between retinalophototrophy and chlorophototrophy.

A) Chlorophototrophic functional units have very high absorption cross sections due to light-gathering antenna pigments (dark green), very high protein mass per functional unit (light green), use iron ions in their internal structure (orange) in addition to protein and organic pigments, and either conserve large amounts of energy per photon or are capable of contributing to biomass production.

B) Functional units in retinalophototrophy have very small absorption cross sections (purple), little protein mass (pink), and conserve low energy per photon.

One of the greatest differences between retinalophototrophs and chlorophototrophs is the efficiency of conversion of light energy into biologically available energy. Chlorophototrophs have a significantly higher energy yield per captured photon than retinalophototrophs. Retinalophototrophic machinery pumps one proton per photon across the cell membrane, while the chlorophototrophic machinery is capable of pumping multiple protons per photon. Most commonly two protons are pumped per photon, via a cytochrome bc proton-pumping complex (related to mitochondrial complex III) or an alternative complex III passing electrons between quinones and cytochromes. Up to four protons per photon is possible for some fraction of electrons in oxygenic phototrophs when a type I reaction center is used with electrons passing from ferredoxin through a complex I-like NDH complex and cytochrome b_6f. However the difference in available energy may be even greater than this ratio would indicate, as microbial rhodopsins are incapable of pumping against membrane polarization of 200mV, which is lower than the proton motive force generated by respiratory electron transport chains. The electron transport chains of chlorophototrophs are thus able to conserve more energy per proton than rhodopsins are by reaching a higher membrane voltage. Furthermore, electrons energized by some chlorophototrophic reaction centers may be passed to electron carriers such as ferredoxin and NADPH, or the high proton-motive force they generate can be used to force reverse electron flow through a respiratory electron transport chain into these carriers for carbon and nitrogen fixation. Retinalophototrophs are unable to produce biomass de novo using their phototrophic machinery, likely due to their low maximum proton-motive force being insufficient to allow reverse electron flow.

The material composition of the phototrophic machinery in retinalophototrophs and chlorophototrophs is also quite different. While microbial rhodopsins consist of a single kDa protein molecule attached to one or two photoactive cofactors per functional unit, chlorophototrophic reaction centers consist of - core protein molecules and a number of accessory proteins per functional unit with a mass of up to kDa, with a large number of diverse photopigments bound to each complex. Moreover, nearly every chlorophototrophic reaction center is associated with multiple diverse antenna complexes, which both greatly increases the absorption cross section per functional unit and can bring the total protein mass per functional unit into the megadaltons or even more. This increased absorption cross section leads to a significantly greater efficiency in terms of captured biological energy per unit incident light at low light intensities for chlorophototrophs, at the expense of saturation at relatively low light levels.

Another relevant difference between chlorophototrophs and retinalophototrophs is the requirement for iron in the chlorophototrophic machinery. While bacterial rhodopsins are entirely composed of protein and organic molecules, every known chlorophototrophic reaction center contains iron atoms. All type I reaction centers contain at least 4 in an Fe-S cluster, with up to 12 in the case of Photosystem I of cyanobacteria and 14 in the case of Acidobacteria and Chlorobi. All type II reaction centers contain one Fe^{+} ion bound at the interface between subunits with additional hemeirons present in Photosystem II of cyanobacteria and an integral cytochrome with additional hemeirons present in the reaction centers of many other lineages. All known electron transport chains that chlorophototrophic reaction centers participate in also utilize iron in their proton pumping components, with iron atoms present in each subunit of cyanobacterial cytochrome bf and the NDH complex used for circular electron flow around Photosystem I containing at least twelve or possibly more. Certain picocyanobacteria reduce their electron transport chains to a form which requires nearly only the iron atoms in the photosystems themselves (biased towards photosystem II with fewer iron atoms) and an alternative oxidase, but this comes at the expense of depressing proton yield to only one proton per photon. This constitutive requirement of iron for functional chlorophototrophy but not retinalophototrophy represents a major resource limitation for chlorophototrophs, especially in oligotrophic environments such as the open ocean where iron levels are limiting.

Ecological Niche Partitioning between Phototrophic Pathways

At first glance the functional differences between retinalophototrophy and chlorophototrophy appear to stem entirely from their disparate evolutionary histories and compositions. However, upon closer inspection these differences appear remarkably coordinated, suggesting that the two of them have precisely partitioned the space of phototrophic ecological niches in two, each filling a different and complementary subset. This has fundamental implications for the early evolution of phototrophy and evolutionary innovations more broadly.

Most trivially, the light-gathering pigments used by the core machinery of chlorophototrophs and retinalophototrophs are spectrally distinct. Retinal primarily absorbs the green wavelengths of visible light, while chlorophyll primarily absorbs in the red and blue wavelengths. This apparent partitioning of the electromagnetic spectrum is somewhat mitigated by the fact that chlorophototrophs contain many accessory pigments aside from basic chlorophyll which can expand their effective absorption spectrum into the green wavelengths.

Nutrient requirements differ substantially between the two systems. Retinalophototrophy solely uses a small protein and an organic pigment to pump protons while chlorophototrophic reaction centers contain iron and are dependent upon functionally-coupled electron transport chain components that also require iron. Retainalotrophy is thus favored under low-iron conditions, which are pervasive throughout much of the oceans. Indeed, it appears that up to % of individual bacterial cells present in the oligotrophic open ocean express microbial rhodopsins, declining in frequency in more nutrient rich environment. This iron-dependent niche partitioning is illustrated well in polar diatoms, eukaryotic phototrophs utilizing both oxygenic photosynthesis and proton-pumping rhodopsins: rhosopsin expression is sharply up regulated during iron starvation in a homeostatic response to maintain energy metabolism. Iron limitation thus favors retinalophototrophy, even though it is significantly less efficient per unit photon absorbed.

The greater efficiency per unit light intercepted by chlorophototrophs compared to retinalophototrophs (along with their capacity to perform redox reactions and thereby directly fix carbon) would, at first glance, imply they are strictly superior under all circumstances without iron limitation. However, efficiency of energy capture per absorbed photon represents just one element of a complex ecological and biophysical trade-off. All metabolic machinery carries with it an investment cost in protein mass - the infrastructure must be built before it can transduce energy or nutrients, and has a finite lifetime before being either recycled or diluted away by growth and division. As such, every metabolic pathway also has a rate of return on investment. In order to determine the return on investment available from multiple phototrophic pathways, one must take into account the mass per functional unit, the rate of operation of the protein machinery, and the yield per cycle, yielding a specific energy flux per unit protein mass.

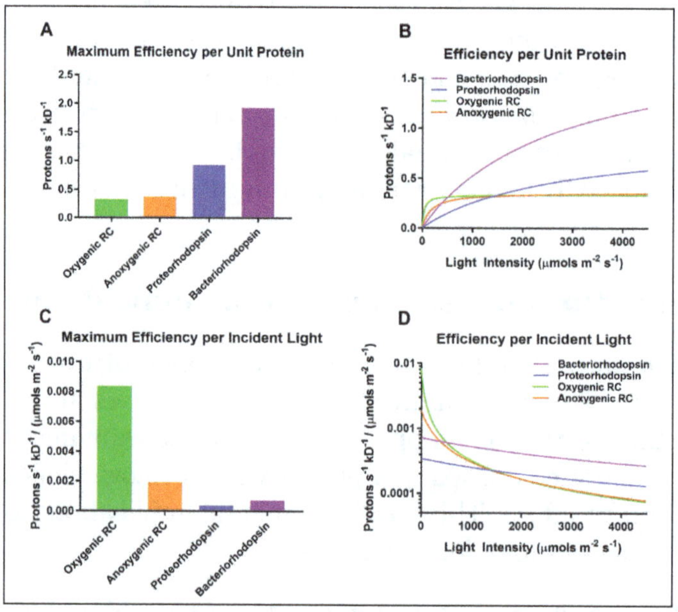

Figure: Ecological comparison between chlorophototrophy and retinalophototrophy.

A) Calculated maximum pumped proton flux available per kDa of protein mass of anoxygenic purple bacterial reaction centers, oxygenic reaction centers, proteorhodopsin, and bacteriorhodopsin at saturating levels of light. Microbial rhodopsins saturate at much higher specific metabolic energy fluxes than reaction centers.

B) Calculated proton flux available per kilodalton of mass of different phototrophic systems at different light intensities. Chlorophototrophic reaction centers produce more energy flux at low light levels compared to microbial rhodopsins, but saturate quickly, while microbial rhodopsins function best at high light levels with higher specific metabolic energy fluxes.

C) Calculated maximum pumped proton flux available per kDa per unit incident light in microeinsteins per square meter. Chlorophototrophic reaction centers are capable of extracting much more energy flux per unit incident light.)

D) Calculated energy flux per kilodalton of machinery per microeinstein of incident light.

Chlorophototrophic reaction centers are significantly more efficient per unit incident light when light is scarce, but rapidly saturate due to having large absorption cross sections per reaction center, thereby gathering more light than can be converted to energy. Microbial rhodopsins, on the other hand, are significantly less efficient per photon, but use light more efficiently than chlorophototrophy when light levels are high.

Table: Chlorophototrophic and Retinalophototrophic Energy Flux per Unit Mass.

	Anoxygenic RC Oxygenic RC	Proteorhodopsin Bacteriorhodopsin
Total Protein mass / RC	~835 kDa ~2098 kDa	27 kDa 26 kDa
Electrons s^{-1}	~150 ~350	0 0
Protons s^{-1}	~300 ~700	~25 >50
Protons s^{-1} kDa^{-1}	0.36 0.33	0.92 >1.92
Normalized Protons s^{-1} kDa^{-1}	1.08 1	2.78 >5.76

We calculated this effective energy flux per unit investment of different phototrophic systems based on a literature review of these values for anoxygenic chlorophototrophic RCs, oxygenic RCs, and two different microbial rhodopsins (proteorhodopsin and bacteriorhodopsin). We quantified the effective flux in terms of protons pumped per kilodalton per second at saturating light levels. Despite their higher efficiency per photon absorbed and faster photocycle, chlorophototrophic machinery is so much more massive than microbial rhodopsins that their specific energy flux per unit mass is significantly lower. Proteorhodopsin and bacteriorhodopsin are calculated as.-fold and.-fold more efficient per unit investment than oxygenic RCs respectively, with anoxygenic RCs roughly equivalent to oxygenic RCs.

A primary reason for low energy flux per unit investment in chlorophototrophic machinery is the presence of large antenna pigments which feed absorbed light energy into reaction centers. This means that the effective absorption cross section per functional chlorophototrophic unit is much larger than the cross section per retinalophototrophic unit. The relative performance of these systems thus varies drastically according to ambient light intensity. When calculating the effect of these differences between absorption cross sections and saturation of the phototrophic machinery at varied light levels, we find that this greater return per unit investment for retinalophototrophs only manifests at high light. The small protein mass and presence of only a single retinal pigment in a microbial rhodopsin ensures a small cross section which requires intense ambient light for the machinery to be used effectively.

Conversely, the large absorption cross section available to the massive chlorophototrophic system is nearly saturated above low light levels of less than microeinsteins per square meter per second, but at these lower light levels maintains a higher energy flux per unit infrastructure. By dividing the function of the return per unit investment of each phototrophic system by the level of ambient light, we produced functions of the efficiency per unit ambient light in units of protons pumped per kilo Dalton per second, per microeinstein of light per square meter per second. While chlorophototrophic reaction centers are more efficient per unit ambient light in the limit of low light, this is reversed at higher light levels.

The differences between chlorophototrophy and retinalophototrophy stem from an intrinsic bio physical trade-off. It is not possible to build a phototrophic system that has both high metabolic efficiency per unit investment (protein infrastructure), and high metabolic efficiency per unit of a rare limiting resource (ambient light). Chlorophototrophy is efficient per unit light at low light levels and requires large amounts of protein investment, while in high light levels retinalophototrophy produces higher energy flux at lower levels of investment. This fits the observed physical distribution of phototrophs in the ocean and ecological distribution of these pathways. Retinalophototrophs are observed at their highest levels in surface ocean waters with high light levels, while chlorophototrophs become most common at slightly deeper levels of the ocean at which light has been partially absorbed. Chlorophototrophy requires a significant fraction of the proteome to be invested to result in an effective energy flux and is the only phototrophic pathway observed in obligate phototrophs, Retinalophototrophy is observed in fully % of bacteria in the open ocean and is frequently present in heterotrophs,,, which appear to frequently use it as a backstop to prevent starvation and increase biomass yield of heterotrophic metabolism.

A similar trade-off is observed across the diversity of heterotrophic metabolic machineries. The difference between respiration and fermentation itself is an example - respiration can produce several times the ATP per unit substrate consumed while producing less than half the energy flux per unit protein mass. The two most common glycolytic pathways - the Entner-Doudoroff (ED) and Embden-Meyerhof-Parnas (EMP) pathway - share precisely this relationship as well. The EMP pathway produces twice the ATP per unit carbohydrate consumed as the ED pathway, but requires -fold more protein mass, and thus produces approximately % the energy flux per kilodalton of protein. Just as chlorophototrophic pathways use more protein then retinalophototrophy to acquire more energy from a small quantity of light and are seen in obligate phototrophs, the EMP pathway is seen more frequently than the ED pathway in obligate anaerobes which cannot switch to aerobic respiration and must obtain more energy from their limited available substrate.

Differences in the cost of metabolic machinery have major implications for growth and ecology. The larger the fraction of a cell's proteome must be put towards the generation and maintenance of energy and resources, the smaller the fraction of the proteome can go towards growth and development. This leads to a series of 'growth laws', which dictate that, all else being equal, a larger investment of protein being used to efficiently consume a rare resource leads to a slower growth rate due to less investment in ribosomes and anabolic functions. The optimal allocation of costly metabolic enzymes under situations of differing growth rate and resource availability therefore explains much of the long-observed trade-off between microbial growth rate and biomass yield. Rapid growth and low yield occurs on abundant resources, while slow growth and high yield occurs on scarce resources. Only recently has optimal proteome allocation been analyzed in the context of phototrophy and autotrophy in general–, but the principles are identical when ambient light is treated as a metabolic resource.

Thus, chlorophototrophy and retinalophototrophy have partitioned phototrophic niche space. Chlorophototrophy is a high-investment strategy suitable for environments of low growth rate, low ambient light resources, or for specialists investing heavily in a single pathway in any environment (i.e., obligate photoautotrophs). Retinalophototrophy is a low-investment strategy suitable for situations of higher growth rate, high ambient light resources, or for flexible metabolic generalists capable of using either phototrophy or heterotrophic metabolism. Taken together with other divergent properties, including chlorophototrophy's requirement of limiting iron and the ease of horizontal transfer of retinalophototrophic capacity over evolutionary time, the properties of these two phototrophic pathways are strikingly complementary.

Nitrogen Fixation

Nitrogen fixation is the essential biological process and the initial stage of the nitrogen cycle. In this process, nitrogen in the atmosphere is converted into ammonia (another form of nitrogen) by certain bacterial species like Rhizobium, Azotobacter, etc. and by other natural phenomena.

Plants are the main sources of food. The nutrients obtained from plants are synthesized by plants using various elements which they obtain from the atmosphere as well as from the soil. This group of elements includes nitrogen as well. Plants obtain nitrogen from the soil and utilise in the process of protein synthesis. Unlike carbon dioxide and oxygen, atmospheric nitrogen cannot be obtained through stomata of leaves. Because the nitrogen gas present in the atmosphere cannot be directly used by plants. There are certain bacteria and some natural phenomenon which help in Nitrogen fixation.

Nitrogen fixation is of two types:

- Non Biological Fixation

Atmospheric nitrogen fixation (or Natural) by lightning:- It contributes about 10% of the total nitrogen fixation. Usually atmospheric nitrogen fixation (or Natural) by lightning occurs in rainy season during lightning or thunder storms.

During lightning the the free nitrogen combines with the oxygen to form nitric oxide.

$$N_2 + O_2 \xrightarrow[\text{Thunder}]{\text{lightning}} 2\,NO \text{ (nitric oxide)}$$

Nitric oxide oxidises into nitrogen dioxide in presence of excess oxygen.

$$2NO + O_2 \xrightarrow{\text{Oxidation}} 2\,NO_2 \text{ (Nitrogen dioxide)}$$

Nitrogen dioxide may react with only water to produce nitric and nitrous acids.

$$2\,NO_2 + H_2O \longrightarrow HNO_3 + HNO_2$$

Or may react with the atmospheric oxygen and rain water to form nitric acid

$$4\,NO_2 + 2\,H_2O + O_2 \longrightarrow 4\,HNO_3 \text{ (Nitric acid)}$$

These acids reaches the soil with rain water and combines with alkaline substances readily release the hydrogen, forming nitrate and nitrite ions.

$$HNO_3 \longrightarrow H^+ + NO_3^- \text{ (nitrate ions)}$$

$$HNO_2 \longrightarrow H^+ + NO_2^- \text{ (nitrite ions)}$$

Figure: The nitrate can be readily utilized by plants and micro-organisms.

- Non Biological Fixation:
    - Industrial fixation through the Haber - Bosch process and combustion. Some ammonia is also produced industrially by the Haber-Bosch process. When nitrogen (dinitrogen) combines with hydrogen in the presence of an iron-based catalyst, at a pressures of 35-100 MPa and fairly high temperature. Usually fossil fuels are used both as a source of energy and hydrogen. Most nitrogenous fertilizers are now derived from atmospheric nitrogen through this type of fixation process.
- Biological fixation by certain microbes — alone or in a symbiotic relationship with some plants and animals:-Biological nitrogen fixation was discovered by the Dutch microbiologist Martinus Beijerinck. It contributes 60% of total nitrogen fixation. But the major conversion of atmospheric N_2 into salts of nitrogen, and then into proteins, is achieved by microorganisms (prokaryotes) such as bacteria, fungi and algae in the process called biological nitrogen fixation (or dinitrogen fixation). Microorganisms that fix nitrogen are called diazotrophs.

Biological nitrogen fixation are of two types:

Free living or non-symbiotic nitrogen fixation: The fixation of free nitrogen of the soil by all the microorganisms living freely or outside the plant cell is called non-symbiotic biological N_2 fixation. It is performed by the aerobic and anaerobic bacteria and blue green algae.

By bacteria Nitrogen fixing bacteria: which are present in the soil convert free nitrogen into soluble compound which are absorbed from the soil by plants. The nitrogen fixing bacteria are of four types:

- Free living non-photosynthetic aerobic nitrogen fixing bacteria e.g., Azotobacter, Beijerinckia and Derxia.
- Free living non-photosynthetic anaerobic nitrogen fixing bacteria e.g. Clostridium.
- Free living photosynthetic nitrogen fixing bacteria e.g,Chromatium, Rhodopseudomonas, Rhodospirillum.
- Free living chemosynthetic nitrogen fixing bacteria e.g., Desulfovibrio.

By free living nitrogen fixing Blue-Green algae: About 15 genera of photosynthetic cyanobacteria (blue-green algae) are found freely in the soil where they fix free N_2 into nitrogenous and ammonium compound. Mostly they are heterocysts e.g., Nostoc, Anabaena, Aulosira, Cylindrospernum, Calothrix. Nitrogen fixation occurs in special thick walled cells termed heterocysts or heterocytes (H) which occur at intervals along the cyanobacterial filaments. This separation of cellular functions is necessary because cyanobacteria have oxygen-evolving photosynthesis but the nitrogen-fixing enzyme, nitrogenase, is unstable in the presence of oxygen. This problem is overcome because the heterocysts contain only part of the photosynthetic apparatus, termed photosystem I, which can be used to generate energy (as ATP). But the heterocysts do not contain photosystem II, which is used to split water into hydrogen (for combination with CO_2 to produce organic products) and oxygen. Non heterocystous nitrogen fixing blue-green algae are less in number e.g., Oscillatoria, Phormidium, Gleocapsa.

Free-living, non-photosynthetic bacteria depend on soil organic matter as a food source whereas the photosynthetic microorganisms may derive their food from the products of photosynthesis.

The nitrogen fixing activity of free-living, non-photosynthetic, aerobic bacteria is strongly dependent on favorable moisture conditions, oxygen, and an organic food source. Anaerobic representatives (Clostridium) predominate in grassland and waterlogged soils and soil aggregates where moisture conditions and organic substrates are available but oxygen supply to the micro-environment of the bacteria is severely restricted.

Symbiotic: Some N2-fixing organisms develops loose (associative) symbiosis with plants or animals (Acetobacter and sugarcane), or establish longer-term relationships within specialized structures provided by their host (Rhizobium and the legume nodule).To provide them with sugars, supplying both a source of energy and a source of carbon for the bacterium's own synthetic reactions. (Symbiosis is a close ecological relationship between the individuals of two (or more) different species. Sometimes a symbiotic relationship benefits both species, sometimes one species benefits at the other's expense, and in other cases neither species benefits).

The fixation of free nitrogen of the soil by N_2-fixing organisms living symbiotically inside plants is known as symbiotic biological nitrogen fixation.

The symbiotic biological nitrogen fixation are of three types:

- Nitrogen fixation through nodule formation in Leguminous plants: The bacteria responsible for the formation of root nodules in leguminous plants belong to the genus Rhizobium. Rhizobium also lives free in soil but only fixes N_2 when inside plant The symbiotic Bacteria Rhizobia (from the Greek words Riza = Root and Bios = Life) are soil bacteria that fix nitrogen (diazotrophy) after becoming established inside root nodules of legumes. According to host specificity and growth of bacteria have been divided into three groups.

Rizobium

Genus	Species	Plant host
Rizobium	leguminosaerum	Peas
Rizobium	Meliloti	Lucerne
Rizobium	Trifolii	Clover
Rizobium	Phaseoli	Beans
Rizobium	lupini	Lupins

- Bradyrhizobium japonicum is the group of slow growing symbiont s of Soybeans (plant host).
- Azorhizobium caulinodans is a bacterium that forms stem nodules in Sesbania (plant host).

The bacteria "invade" the plant and cause the formation of a nodule by inducing localised proliferation of the plant host cells. Root nodules acts as a site of Nitrogen fixation. The root nodules contain a pigment called leghaemoglobin(serving the same function as the oxygen-carrying haemoglobin in blood).The heme (oxygen-binding) portion is produced by the bacterium, while the globin (protein) portion is produced by the host plant, again showing the closeness of the symbiotic relationship. The function of this molecule in nodules is to reduce the amount of free oxygen, and thereby to protect the nitrogen-fixing enzyme nitrogenase, which functions only under anaerobic conditions. Nitrogenase is the only enzyme that can split nitrogen molecule for nitrogen fixation.

- Nitrogen fixation through nodule formation in None-Leguminous plants: There are many plants belonging non- Leguminosae families, specially shrubs and plants which produce root nodules. Example:
    - Frankia is a genus of the bacterial group termed actinomycetes - filamentous bacteria.
    - Frankia form nitrogen-fixing root nodules (sometimes called actinorhizae) with numerous genera of non-leguminous angiosperms, such as alder (Alnus species), sea buckthorn (Hippophae rhamnoides, which is common in sand-dune environments) and Casuarina (a Mediterranean tree genus).
    - Alder and the other woody hosts of Frankia are typical species that invade nutrient-poor soils. These plants probably benefit from the nitrogen-fixing association, while supplying the bacterial symbiont with photosynthetic products.
    - Rhizobium also forms nitrogen-fixing root nodules with genus Parasponia.
    - Sometimes nodules are also formed in the roots of certain gymnosperms e.g., Podocarpus and in the leaves of Pavetta zinumermanniana and Chomelia.
- Nitrogen fixation through non-nodulation: In some plants root nodules are not formed but symbiotic nitrogen fixation takes place. Examples:
    - Lichens live as symbionts with photosynthetic cyanobacteria (blue green algae or Green chlorophyllous and with fungi.
    - Anthroceros (Bryophte): It contains Blue green alga Nostoc inside mucilage cavities present on ventral side.
    - Azolla: The water fern, Azolla lives symbiotically with the nitrogen-fixing cyanobacteria (Anabaena azollae) Azolla is grown in rice paddies early in the season. As the rice grows above the water surface, it shades out the fern, which dies, releasing the stored nitrogen. In this way, the paddy is fertilized without application of chemical fertilizers.
    - Cycas (gymnosperms): It contains cyanobacteria (blue-green algae) Anabaena or Nostoc. Aerial roots contain a nitrogen-fixing cyanobacterial symbiont.
    - Gunnera macrophylla (angiosperms): Its stem contains Nostoc.

Associative Symbiotic Nitrogen Fixation: When bacteria form a close association with the roots of cereals and grasses and fix nitrogen, the association is of loose mutualism type and known as loose (associative) symbiosis and this type of nitrogen fixation is known as associative symbiotic nitrogen fixation. The bacteria grow in the rhizosphere in close contact with the roots, sometimes invade the outer cortical regions of the roots, and fix nitrogen. Azospirillum bransilense (= Spirillum lipoferum) a bacterium discovered J. Dobereiner, is the bacterium forming associate-symbiosis with the cereal roots. Others are Pseudomonas azotogensis, Enterobacter, Bacillus, Klebsiella etc.

Rates of symbiotic N_2 fixation in legumes vary with plant species and cultivator, growing season, and soil fertility. Some forage legumes can fix 600 kilograms per hectare per year but more common values are 100 to 300 kilograms per hectare per year. Rates for grain legumes are often lower.

Inclusion of legumes in crop rotations is generally thought to improve soil nitrogen levels, but benefits depend on the level of N2 fixed and the amount of nitrogen removed in grain or forage. A good soybean crop might fix 180 kilograms per hectare but remove 210 kilograms per hectare in the grain. Non-symbiotic bacteria fix only up to 5 kg of nitrogen per hectare in one year <pFormation of Nodule:

- Rhizobia is the group of genera of alpha-proteobacteria (family Rhizobiaceae) which includes all of the nitrogen-fixing species that produce nodules with legumes such as clover and soybean, Allorhizobium, Azorhizobium, Bradyrhizobium, Mesorhizobium, Phyllobacterium, Rhizobium, and Sinorhizobium, as well as the plant pathogen Agrobacterium.
- Rhizobia produce stem or root nodules on their host(s), and within these nodules receive protection from external stresses and energy for growth and N_2 fixation. The host receives most of the nitrogen it needs for growth.
- Only infection via root hairs is considered here.
- Plants of legume family secrete flavonoids which are recognized by bacterial NodD protein when NodD binds flavonoid it activates other nodulation genes.
- Rhizobium secretes Nod factors (Some nod genes encode enzymes that make Nod factors) the leguminous plant recognizes Nod factors.
- Then growing root hair of a leguminous plant comes in contact with the bacterium -Rhizobium, the growing root hairs curl and form a pocket for the rhizobia. the bacteria invade the plant by a newly formed infection thread growing through it. the root hair cell wall.
- Simultaneously, cortical cells are mitotically activated, giving rise to the nodule primordi-um.
- Infection threads grow toward the primordium, and the bacteria are then released into the cytoplasm of the host cells, surrounded by a plant derived peribacteroid membrane (PBM). This separation is usually seen as a mechanism to suppress plant defense responses likely to harm the bacteria.

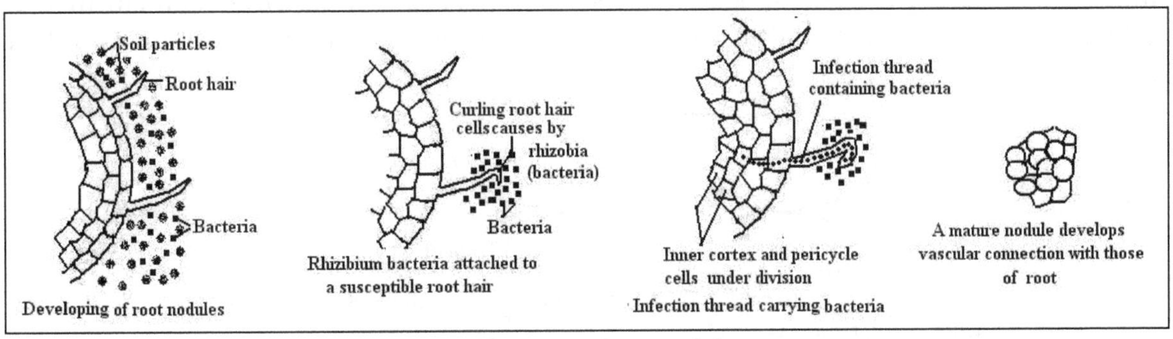

Figure: Development of root nodule.

- With the production of the infection threads, bacteria produce cytokinins (type of plant hormone). Cytokinins promote division of plant cells to form nodules and nodules begin to form in the root hairs.
- The nodule primordium thereupon develops into a mature nodule, while the bacteria

differentiate into their endosymbiotic form, which is known as the bacteroid. Bacteroids, together with the surrounding PBMs, are called symbiosomes.

- Cell division now sets in, in the infected tissue leading to nodule formation. The area of active N_2 fixation is either pink or red in color due to the presence of leghaemoglobin needed for oxygen transport. The nodule thus formed establishes a direct vascular connection with the host for the exchange of nutrients.
- All of the steps of nodule development involve the expression of nodule-specific plant genes, the so called nodulin genes. The early nodulin genes encode products that are expressed before the onset of nitrogen fixation and are involved in infection and nodule development. The products of the late nodulin genes are involved in the interaction with the endosymbiont and in the metabolic specialization of the nodule.

Mechanism of Nitrogen Fixation

The bacteroids within the nodules formed on legume roots fix nitrogen.

Structure and Operation of Nitrogenase

The nitrogenase is an enzymatic complex, which converts atmospheric nitrogen (N_2) to ammonia. The nitrogenase complex exists in both free living nitrogen fixing organisms as well as in symbiotic nitrogen fixing bacteria. Nitrogenase is a complex of two separately isolated proteins- an iron protein or dinitrogen reductase and a molybdenum-iron protein or dinitrogenase. The proteins have a negative redox potential. The MoFe protein (Iron-Molybdenum protein), is a heterotetramer composed of two alpha subunits and two beta subunits. The protein contains two copies of each of two types of clusters: P clusters and FeMo cofactors. Each P cluster contains 8 iron atoms and 7 sulfides linked to the protein by 6 cysteinate residues. Each FeMo cofactor contains one molybdenum atom, 7 iron atoms, and 9 sulfides. This protein is responsible for reducing atmospheric nitrogen to ammonia via a series of electron transfers within the protein to the substrate molecule. The reaction requires the addition of six electrons for each nitrogen molecule that is split into two ammonia molecules. The Fe protein (Nitrogenase Reductase -NR) is a dimer and formed by 2 subunits of polypeptide chains linked by a 4Fe-4S cluster. Each monomer contains an ATP binding site.

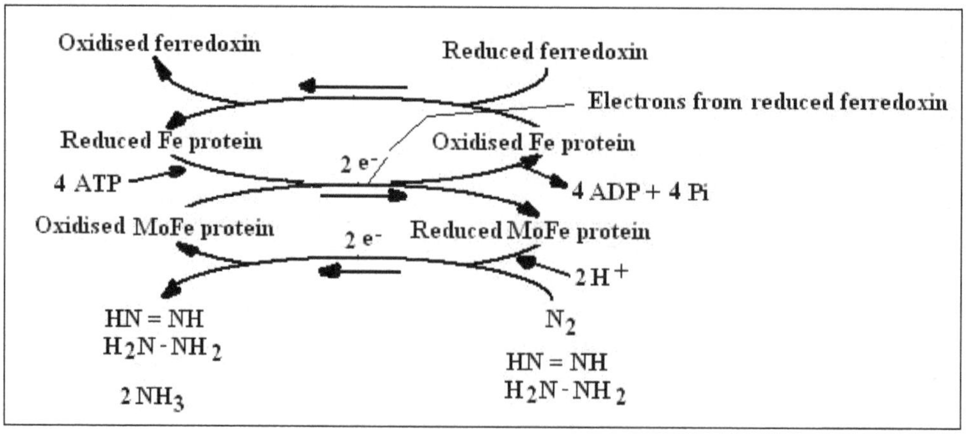

Figure: Working of Nitrogenase.

The Fe protein is first reduced by electrons donated by ferredoxin. Then the reduced Fe protein binds ATP and reduces the molybdenum-iron protein, which donates electrons to N_2, producing HN=NH. In two further cycles of this process (each requiring electrons donated by ferredoxin) HN=NH is reduced to H_2N-NH_2, and this in turn is reduced to $2NH_3$.The reactions occur while N_2 is bound to the nitrogenase enzyme complex.

The favorable condition for nitrogen fixation is:

- Presence of enzymes nitrogenase and hydrogenase in the nitrogen fixing cells or organisms.
- Presence of leghaemoglobin which protect the enzyme nitrogenase from oxygen.
- Ferredoxin which supplies electrons for this process.
- A source of hydrogen (strong reducing agent) like NADPH or $FMNH_2$.
- Constant supply of ATP to transfer hydrogen atoms to dinitrogen. ATP is provided by aerobic respiration of sugars, ultimately produced by photosynthesis. Phosphorous is an important component of the biochemical energy source, ATP (adenosine triphosphate). Thus, for legumes to fix N, there must be adequate available soil P.
- Presence of coenzymes and cofactors.
- Compounds for trapping ammonia formed by the reduction of dinitrogen (N_2).

Nitrogen fixation is controlled by plant nod genes and bacterial nod, nif and fix gene cluster. Biological nitrogen fixation by free living and symbiotic bacteria is carried out by step by step progressive reduction of dinitrogen (N_2) molecules by the addition of a pair of hydrogen atoms. Depending on the type of microorganism, the reduced ferredoxin which supplies electrons for this process is generated by photosynthesis, respiration or fermentation. In the heterocystous bacteria the primary electron donor to nitrogenase is also a ferrredoxin, but it receives electrons produced by the action of light on the photosynthetic apparatus. The electrons are supplied via ferredoxin to nitrogenase reductase and then nitrogenase. The reductase donates 8 electrons in succession to the nitrogenase cofactor, a molybdenum-iron containing active center which catalyzes the actual reduction of dinitrogen. Iron (Fe^{+3}) and molybdenum (Mo^{+4}) of enzyme nitrogenase takes part in attachment of a dinitrogen molecule (N_2) and weaken the bonds between the two atoms of the nitrogen. The weakened molecule of nitrogen is reduced by the reducing agent (NADPH,$FMNH_2$).

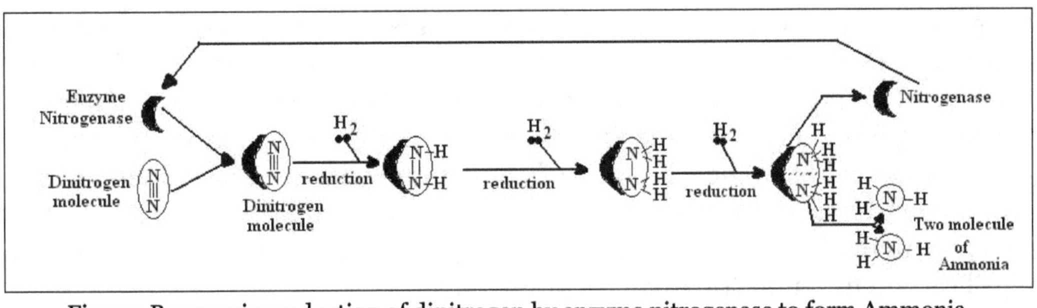

Figure: Progressive reduction of dinitrogen by enzyme nitrogenase to form Ammonia.

- It produces dimide (N2H2), hydrazine (N2H4) and then ammonia (2NH3)Where one molecule of N2 is reduced in the presence of protons to 2 NH3, and H2 as a byproduct. Semi-activated nitrogenase can reduce easy substrates such as acetylene. In the typical reaction, two molecules of ATP are consumed for each electron transferred.16 ATP's are needed to fix a single nitrogen molecule in nitrogen fixation, the plant regulates the nitrogenase's activity and expression according to reduced nitrogen availability and oxygen presence. Molecular oxygen is a strong inhibitor of the nitrogenase Mo-Fe cofactor and is removed by the plant oxygen binding protein leghemoglobin in the root nodules.Some of the cyanobacteria have yet another mechanism for protecting nitrogenase: nitrogen fixation occurs in special cells (heterocysts) which possess only photosystem I (used to generate ATP by light-mediated reactions) whereas the other cells have both photosystem I and photosystem II (which generates oxygen when light energy is used to split water to supply H2 for synthesis of organic compounds). Nitrogenase also converts hydrogen ions to hydrogen gas at the same time thus consuming even more ATP in the process.

Nitrogen Metabolism

All of the nitrogen in a plant, whether derived initially from nitrate, nitrogen fixation, or ammonium ions, is converted to ammonia, which is rapidly incorporated into organic compounds through a number of metabolic pathways is known as nitrogen metabolism,consists of:

- Nitrogen fixation.
- Nitrogen reduction or Nitrification: Ammonia formed as a result of nitrogen fixation is used for the synthesis of amino acids. Amino acids are transported through phloem to other parts for the synthesis of proteins.Ammonium ions can be taken by higher plants but plants are more adapted to absorb nitrate (NO^{3-}) than ammonium ions (NH^{4+}) from soil. Nitrification is an aerobic microbial process by which specialized bacteria oxidize ammonium to nitrite and then to nitrate. It is accomplished by nitrifying bacteria like nitrosomonas, nitrosococcus and Nitrobacter. Nitrification is a two-step process. The first stage is the oxidation of ammonium (NH^{4+}) to nitrite (NO^{2-}), a function carried out by bacteria in the genus Nitrosomonas. The nitrite formed is rapidly oxidized to nitrate (NO^{3-}) by bacteria in the genus Nitrobacter.

$$\underset{\text{(Ammonia)}}{NH_4^+} + 2O_2 \xrightarrow[\text{Nitrosococcus}]{\text{Nitrosomonas}} \underset{\text{(Nitrite ion)}}{NO_2^-} + 2H_2O + \text{Energy}$$

$$\underset{\text{(Nitrite ion)}}{NO_2^-} + O_2 \xrightarrow{\text{Nitrobacter}} \underset{\text{(Nitrate ion)}}{NO_3^-} + \text{Energy}$$

The nitrifying bacteria nitrosomonas, nitrosococcus and Nitrobacter are chemoautotrophs: They gain their energy by chemical oxidations (chemo-) and they are autotrophs (self-feeders) because they do not depend on pre-formed organic matter. As they derive energy for synthesis of organic food by oxidizing inorganic materials ammonia. Nitrification is an autotrophic process during which energy is liberated from the oxidation of ammonium with the biosynthesis of simple inorganic molecules such as carbon dioxide and water into organic compounds and oxygen is an electron acceptor. nitrifying bacteria, gain their energy by oxidizing ammonium, while using CO_2 as their source of carbon to synthesize organic compounds

Nitrogen assimilation: Nitrogen assimilation is the conversion of inorganic nitrogen (such as nitrate) into an organic form of nitrogen like, for example, an amino acid. Nitrate is reduced for this purpose by enzymes first to nitrite (by nitrate reductase), then to ammonia (by nitrite reductase).

Ammonia is incorporated into amino acids. The process of nitrate reduction to ammonia is accomplished in two steps, each mediated by specific enzymes:

- Reduction of nitrate to nitrite: The nitrate serves as a terminal electron acceptor for anaerobic respiration. The nitrate is reduced to nitrite by enzyme nitrate reductase. It is co-enzyme NADH/NADPH-dependent according to organism.

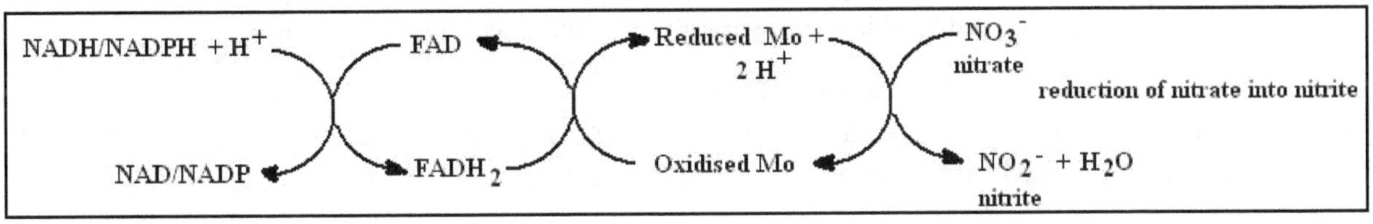

- The enzyme is a flavoprotein that contains iron and molybdenum serves as an electron carrier. FAD receives hydrogen from reduced co-enzyme NADH/NADPH +H+ (serves as hydrogen donor) for the reduction of nitrate.

- Reduction of nitrite to ammonia: Nitrite reductase reduces the nitrite ions to ammonium ions. Nitrite reductase does not require molybdenum and may contain copper and iron. Ferredoxin is the direct source of electrons for nitrite reduction, which occurs in higher plants mostly in the leaves.

$$2NO_2 + 7NADPH + 7H^+ \xrightarrow[\text{Ferredoxin}]{\text{Nitrite reductase}} 2NH_3 + 4H_2O + 7NADP^+$$

The nitrite ions formed in other parts of the plant are also transported to leaves and reduced to ammonia. The reduced coenzyme NADPH +H+ or NAD + H+ serves as hydrogen donor for the reduction of nitrite.

Ammonia thus formed as a result of nitrogen fixation is not given out.It is highly toxic and used for the synthesis of amino acids. Amino acids are the building blocks for the synthesis of proteins. The amino acids are transported through phloem to other parts of the plant for the synthesis of proteins.

Amino acids are the initial products of nitrogen assimilation. An amino acid molecule consists of at least one carboxyl (-COOH) groups and one or several amino (-NH_2) groups. Majority of amino acids are synthesized in plants by two main processes:

Reductive animation: In this process, ammonia reacts with alpha-ketoglutaric acid to form glutamic acid in the presence of enzyme glutamate dehydrogenase. A reduced coenzyme NADPH in leaves, NADH in roots is required.

$$NH4^+ + \alpha - Ketoglutaric + NADPH \xrightarrow[\text{dehydrogenase}]{\text{Enzyme: Glutamate}} Glutamate + H_2O + NADP$$

Transamination: Glutamic acid is the main amino acid from which other 17 amino acids are formed through transamination. With the help of enzyme transaminase, amino group of an amino acid (-CHNH2) is exchanged with keto group (-CO) of keto acid. Pyridoxal phosphate is required as coenzyme which is obtained from a vitamin.

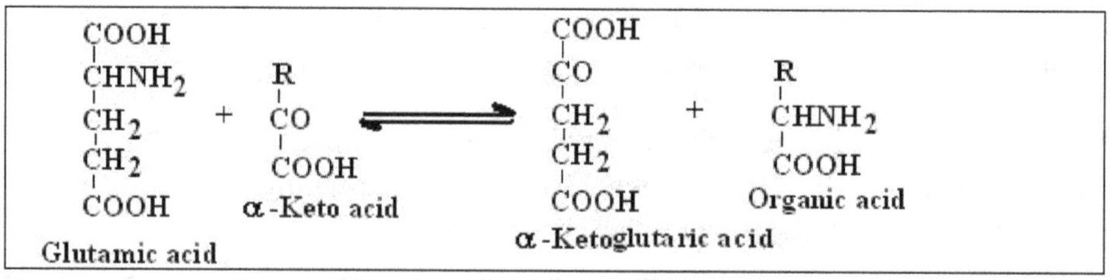

A very common a-ketoacid is a-ketoglutarate, an intermediate in the citric acid cycle.

In most of transamination glutamic acid is present as one of the reactants. The other reactant may be any one of a number of alpha-keto acids,. Which alpha keto acid receives the amino group from glutamic acid is determined by the specificity of the enzyme.

Amides

Amides are nitrogen-containing organic compounds and contain more nitrogen than amino acids. Amides are formed when amino acids bond to each other to form proteins. Amino acids in which hydroxyl group (-OH) of carboxylic group is replaced by amino group (-NH_2) from ammonia, or another amino acid. ATP is required. The two important amides found in plants are asparagine and glutamine.

$$NH4^{+} + Aspartic\,acid + ATP \xrightarrow[synthetaste]{\substack{Enzyme:\\ Asparagine}} Asparagine + ADP + Pi$$

$$Glutamine + Aspartic\,acid + ATP \xrightarrow[transamidase]{\substack{Enzyme:\\ Glutamine-aspartate}} Glutamic\,acid + Asparagine + ADP + Pi$$

They are formed from two amino acids namely glutamic acid and aspartic acid. This reaction takes place in the presence of the enzymes glutamine synthetase or asparagine synthetase.

Proteins synthesis: Proteins are made up of long chains of amino acids. Proteins are in the form of one or more chain called polypeptide chains. Amino acids bond to each other by peptide or amide bonds. The carboxyl group (-COOH) of one amino acid reacts with the amino group (-NH_2) of the next amino acid, releasing a molecule of water and as a result peptide bond (-CONH_) is formed. This may be illustrated with the two simplest amino acids, glycine and alanine.

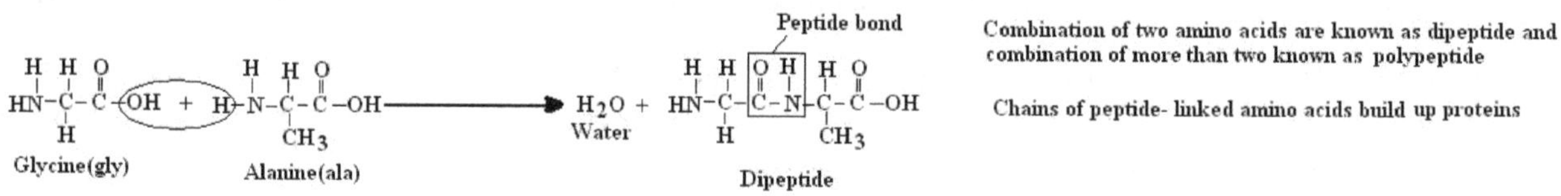

Figure: In a polypeptide, amino acids are arranged according to coded information contained in mRNA.

Polypeptide synthesis occurs over ribosome where mRNA gets attached. Amino acids are brought there according to codons by means by tRNAs. The number of amino acids varies greatly among proteins and thus differs from the molecular weight of proteins also.

References

- Microbial-metabolism, metabolism, 5-major-metabolic-pathways-in-organisms-microbiology-65509: biologydiscussion.com, Retrieved 05, May 2020
- Energy-matter-and-enzymes, microbiology: courses.lumenlearning.com, Retrieved 10, June 2020
- Chemolithotrophy, boundless-microbiology: courses.lumenlearning.com, Retrieved 24, April 2020
- What-is-Anabolism, life-sciences: news-medical.net, Retrieved 20, January 2020

Chapter 3

Microbial Genetics

Microbial genetics can be defined as the study of genes and gene functions in microorganisms for various purposes. Microbes have relatively simple genetic systems and their study is crucial to the development of molecular and cell biology among others. Bacterial genetics, archaeal genetics, fungal genetics, etc. are some of its aspects which are discussed in detail in this chapter.

Microbial genetics is a branch of genetics concerned with the transmission of hereditary characters in microorganisms. Microorganisms have the ability to acquire genes and thereby undergo the process of recombination. In recombination, a new chromosome with a genotype different from that of the parent results from the combination of genetic material from two organisms. This new arrangement of genes is usually accompanied by new chemical or physical properties.

In microorganisms, several kinds of recombination are known to occur. The most common form is general recombination, which usually involves a reciprocal exchange of DNA between a pair of DNA sequences. It occurs anywhere on the microbial chromosome and is typified by the exchanges occurring in bacterial transformation, bacterial recombination, and bacterial transduction.

A second type of recombination, called site-specific recombination, involves the integration of a viral genome into the bacterial chromosome. A third type is replicative recombination, which is due to the movement of genetic elements as they switch position from one place on the chromosome to another.

The principles of recombination apply to prokaryotic microorganisms but not to eukaryotic microorganisms. Eukaryotes exhibit a complete sexual life cycle, including meiosis. In this process, new combinations of a particular gene form during the process of crossing over. This process occurs between homologous chromosomes and is not seen in bacteria, where only a single chromosome exists. Much of the work in microbial genetics has been performed with bacteria, and the unique features of microbial genetics are usually those associated with prokaryotes such as bacteria.

Bacterial Genetics

Bacterial genetics is a scientific field concerned with the mechanisms of heredity in bacteria. Genetics is the study of genes including the structure of genetic materials, what information is stored in the genes, how the genes are expressed and how the genetic information is transferred. Genetics is also the study of heredity and variation. The arrangement of genes within organisms is its genotype and the physical characteristics an organism based on its genotype and the interaction with its environment, make up its phenotype. The order of DNA bases constitutes the bacterium's genotype. A particular organism may possess alternate forms of some genes. Such alternate forms of genes are referred to as alleles. The cell's genome is stored in chromosomes, which are chains of double stranded DNA. Genes are sequences of nucleotides within DNA that code for functional proteins. The genetic material of bacteria and plasmids is DNA. The two essential functions of genetic material are replication and expression.

Structure of DNA

The DNA molecule is composed of two chains of nucleotides wound around each other in the form of "double helix". Double-stranded DNA is helical, and the two strands in the helix are antiparallel. The backbone of each strand comprises of repeating units of deoxyribose and phosphate residue. Attached to the deoxyribose is purine (AG) or pyrimidine (CT) base. Nucleic acids are large polymers consisting of repeating nucleotide units. Each nucleotide contains one phosphate group, one deoxyribose sugar, and one purine or pyrimidine base. In DNA the sugar is deoxyribose; in RNA the sugar is ribose. The double helix is stabilized by hydrogen bonds between purine and pyrimidine bases on the opposite strands. A on one strand pairs by two hydrogen bonds with T on the opposite strand, or G pairs by three hydrogen bonds with C. The two strands of double-helical DNA are, therefore complementary. Because of complementarity, double-stranded DNA contains equimolar amounts of purines (A + G) and pyrimidines (T + C), with A equal to T and G equal to C, but the mole fraction of G + C in DNA varies widely among different bacteria. One of the differences between DNA and RNA is that RNA contains uracil instead of the base thymine.

Structure of Chromosome

In contrast to the linear chromosomes found in eukaryotic cells, most bacteria have single, covalently closed, circular chromosomes. Not all bacteria have a single circular chromosome: some bacteria have multiple circular chromosomes, and many bacteria have linear chromosomes and linear plasmids. Multiple chromosomes have also been found in many other bacteria, including Brucella, Leptospira interrogans, Burkholderia and Vibrio cholerae. Borrelia and Streptomyces have linear chromosomes and most strains contain both linear and circular plasmids. The chromosome of E coli has a length of approximately 1.35 mm, several hundred times longer than the bacterial cell, but the circular DNA is then looped and supercoiled to allow the chromosome to fit into the small space inside the cell.

Codon

A set of three base pairs constitutes a codon, which codes for a single amino acid. The "triplet code" is said to be degenerate or redundant because more than codon may exist for the same amino acid. For example, the codons AGA, AGG, CGU, CGC, CGA and CGG all code for arginine. There are 64 codons, of which 3 (UAA, UAG and UGA) are nonsense codons. They don't code for any amino acid, but act as stop codons. There are specific codons which code for start and stop sequences. The start codon (AUG) indicates the beginning of the sequence to be translated, and the stop codons (UAA, UGA, UAG) terminate the protein synthesis. With the exception of methionine, all amino acids are coded for by more than one codon. The DNA in a gene that are expressed into the protein product are called exons and the non-coding DNA segments are called introns. There are no introns in bacterial chromosome. A segment of DNA carrying codons specifying a particular polypeptide is called a cistron or a gene.

Flow of Genetic Information

The central dogma of molecular biology is that DNA carries all genetic information. The flow of genetic information includes the replication of DNA to make more DNA, the transcription of the DNA into mRNA and the translation of mRNA into proteins. Replication of DNA first involves the separation of the two strands of DNA followed by synthesis of new identical DNA strand by enzymes called DNA polymerases. The RNA strand is synthesized by enzymes called RNA polymerases. The RNA sequence will be complementary to the DNA sequence. The mRNA strands are then guided to the ribosomes for protein translation. Amino acid residues are brought to the mRNA strand on the ribosomes by transfer RNA (tRNA).

Operon Concept

Some proteins (or enzymes) are always required by a bacterium, genes coding for such proteins are constitutively expressed. These genes are usually needed for the cell to survive. Other genes that may not be needed at all times are regulated to conserve energy and cellular materials. Some proteins (or enzymes) are produced only when the need arises or when stimulated by certain environmental conditions. Such genes are normally repressed and are induced whenever required. Repression is a method of inhibiting or decreasing the expression of specific genetic products. This inhibition is controlled by proteins called repressors, which usually block the binding of RNA polymerase to the template DNA. Induction is the opposite of repression, inducers act to "turn on" genes that are not constitutive. The operon concept was first demonstrated by Jacob and Monad.

Bacteria utilize a special energy saving system of genetic control called operons. The operon is a sequence of DNA that contains multiple genes used to produce multiple proteins for a single purpose. An example of an operon is the lac operon in E. coli. In order to break down lactose, E. coli must use a series of enzymes (beta-galactosidase, galactoside permease and transacetylase). The genes for these three enzymes are located in a row on the DNA and share a single promoter. Genes determining structure of a particular protein are called structural genes and the activity of structural genes are controlled by regulator genes, which lie adjacent to them. The genes lacZ, lacY and lacA which code for the three enzymes are the structural genes. lacI gene codes for the repressor protein, hence is the regulator gene. Between the lacI gene and the structural genes lie promoter and operator genes. For transcription of the structural genes, the enzyme RNA polymerase first has to bind to promoter region. The operator region lies in between the promoter and structural genes and the RNA polymerase has to go through the operator region. Under normal circumstances, when the structural genes are not transcribed, the repressor protein is bound to the operator region thus preventing the passage of RNA polymerase from the operator region towards the operon. When lactose is available in the environment, the repressor protein leaves the operator region and binds to lactose because it has high affinity for lactose. This frees the operator region and the RNA polymerase enzyme moves towards the operon and transcribes the structural genes. The products of structural genes result in the metabolism of lactose. When lactose is no more available, the repressor protein goes back and binds to the operator region, thus stopping further transcription of structural genes. This way lactose acts both as inducer as well as a substrate for beta galactosidase.

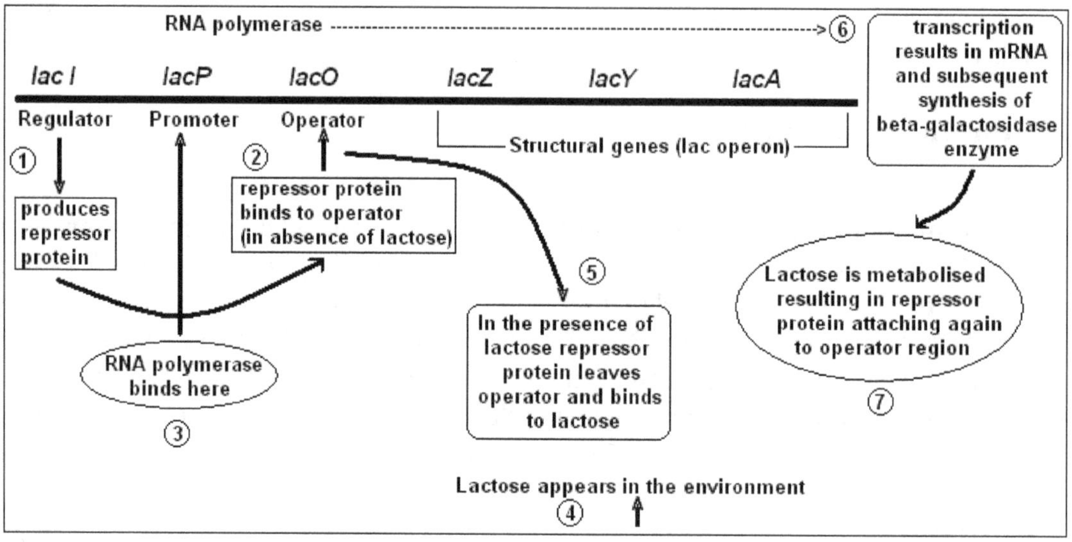

Mutations

The term "mutation" was coined by Hugo de Vries, which is derived from Latin word meaning "to change". Mutations are heritable changes in genotype that can occur spontaneously or be induced by chemical or physical treatments. (Organisms selected as reference strains are called wild type, and their progeny with mutations are called mutants.) The process of mutation is called mutagenesis and the agent inducing mutations is called mutagen. Changes in the sequence of template DNA (mutations) can drastically affect the type of protein end product produced. For a particular bacterial strain under defined growth conditions, the mutation rate for any specific gene is constant and is expressed as the probability of mutation per cell division. Spontaneous mutation occurs naturally about one in every million to one in every billion divisions. Mutation rates of individual genes in bacteria range from 10-2 to 10-10 per bacterium per division. Most spontaneous mutations occur during DNA replication.

Mechanisms of mutation:

- Substitution of a nucleotide: Base substitution, also called point mutation, involves the changing of single base in the DNA sequence. This mistake is copied during replication to produce a permanent change. If one purine [A or G] or pyrimidine [C or T] is replaced by the other, the substitution is called a transition. If a purine is replaced by a pyrimidine or vice-versa, the substitution is called a transversion. This is the most common mechanism of mutation.
- Deletion or addition of a nucleotide: deletion or addition of a nucleotide during DNA replication. When a transposon (jumping gene) inserts itself into a gene, it leads to disruption of gene and is called insertional mutation.

Results of mutation:

- Missense mutation: Missense mutations are DNA mutations which lead to changes in the amino acid sequence (one wrong codon and one wrong amino acid) of the protein product. This could be caused by a single point mutation or a series of mutations.

- Nonsense mutation: A mutation that leads to the formation of a stop codon is called a nonsense mutation. Since these codon cause the termination of protein synthesis, a nonsense mutation leads to incomplete protein products.
- Silent mutation: Sometimes a single substitution mutation change in the DNA base sequence results in a new codon still coding for the same amino acid. Since there is no change in the product, such mutations are called silent.
- Frame shift mutation: Frame shift mutations involve the addition or deletion of base pairs causing a shift in the "reading frame" of the gene. This causes a reading frame shift and all of the codons and all of the amino acids after that mutation are usually wrong. Since the addition of amino acids to the protein chain is determined by the three base codons, when the overall sequence of the gene is altered, the amino acid sequence may be altered as well.
- Lethal mutation: Sometimes some mutations affect vital functions and the bacterial cell become nonviable. Hence those mutations that can kill the cell are called lethal mutation.
- Suppressor mutation: It is a reversal of a mutant phenotype by another mutation at a position on the DNA distinct from that of original mutation. True reversion or back mutation results in reversion of a mutant to original form, which occurs as a result of mutation occurring at the same spot once again.
- Conditional lethal mutation: Sometimes a mutation may affect an organism in such a way that the mutant can survive only in certain environmental condition. Example; a temperature sensitive mutant can survive at permissive temperature of 35 °C but not at restrictive temperature of 39 °C.
- Inversion mutation: If a segment of DNA is removed and reinserted in a reverse direction, it is called inversion mutation.

Based on extent of base pair changes, mutations can be of two types; micro lesion and macro lesion. Micro lesions are basically point mutations (affecting single base pairs) whereas macro lesions involve addition, deletion, inversion or duplication of several base pairs.

The mutations in DNA can occur spontaneously or can be caused by an external force or substance called a mutagen. Mutagens can be chemicals such as nitrous acid, which alters adenine to pair with cytosine instead of thymine. Other chemical mutagens include acridine dyes, nucleoside analogs that are similar in structure to nitrogenous bases, benzpyrene (from smoke and soot) and aflatoxin. Radiation can also be a cause of DNA mutations. High energy light waves such as X-rays, gamma rays, and ultraviolet light have been shown to damage DNA. UV light is responsible for the formation of thymine dimers in which covalent links are established between the thymine molecules. These links change the physical shape of the DNA preventing transcription and replication.

Significance of mutation:

- Discovery of a mutation in a gene can help in identifying the function of that gene.
- Mutations can be induced at a desired region to create a suitable mutant, especially to produce vaccines.
- Spontaneous mutations can result in emergence of antibiotic resistance in bacteria.

- Mutations can result in change in phenotype such as appearance of novel surface antigen, alternation in physiological properties, and change in colony morphology, nutritional requirements, biochemical reactions, growth characteristics, virulence and host range.

Tests to Detect or Select Mutations

- Replica plating
- Penicillin enrichment
- Fluctuation test
- Ames test

Transfer of Genetic Material

Sometimes when two pieces of DNA come into contact with each other, sections of each DNA strand will be exchanged. This is usually done through a process called crossing over in which the DNA breaks and is attached on the other DNA strand leading to the transfer of genes and possibly the formation of new genes. Genetic recombination is the transfer of DNA from one organism to another. The transferred donor DNA may then be integrated into the recipient's nucleoid by various mechanisms. In the case of homologous recombination, homologous DNA sequences having nearly the same nucleotide sequences are exchanged by means of breakage and reunion of paired DNA segments. Genetic information can be transferred from organism to organism through vertical transfer (from a parent to offspring) or through horizontal transfer methods such as conjugation, transformation or transduction. Bacterial genes are usually transferred to members of the same species but occasionally transfer to other species can also occur.

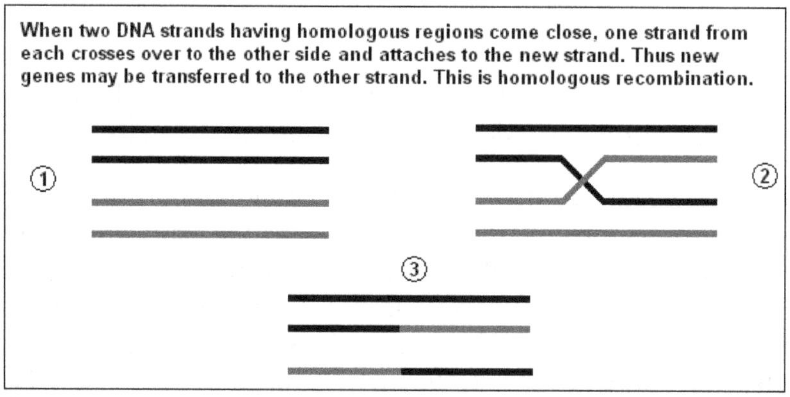

Figure: General or homologous recombination requires extensive homology and is mediated by an enzyme, RecA protein.

Transformation

Transformation involves the uptake of free or naked DNA released by donor by a recipient. It was the first example of genetic exchange in bacteria to have been discovered. This was first demonstrated in an experiment conducted by Griffith in 1928. The presence of a capsule around some strains of pneumococci gives the colonies a glistening, smooth (S) appearance while pneumococci lacking capsules have produce rough (R) colonies. Strains of pneumococci with a capsule (type I) are virulent and can kill a mouse whereas strains lacking it (type II) are harmless. Griffith found that mice died when they were injected with a mixture of live non capsulated (R, type II) strains and heat killed capsulated (S, type I) strains. Neither of these two when injected alone could kill the mice, only the mixture of two proved fatal. Live S strains with capsule were isolated from the blood of the animal suggesting that some factor from the dead S cells converted the R strains into S type. The factor that transformed the other strain was found to be DNA by Avery, McLeod and McCarty in 1944.

Transformation is gene transfer resulting from the uptake by a recipient cell of naked DNA from a donor cell. Certain bacteria (e.g. Bacillus, Haemophilus, Neisseria, and Pneumococcus) can take up DNA from the environment and the DNA that is taken up can be incorporated into the recipient's chromosome.

Rough colonies

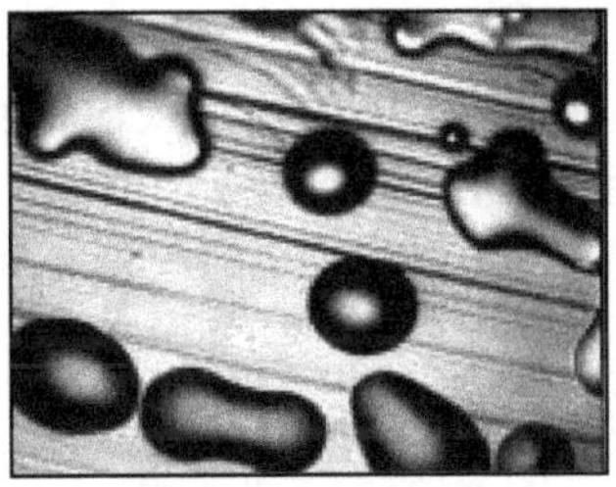

Smooth colonies

The steps involved in transformation are:

- A donor bacterium dies and is degraded.
- A fragment of DNA (usually about 20 genes long) from the dead donor bacterium binds to DNA binding proteins on the cell wall of a competent, living recipient bacterium.
- Nuclease enzymes then cut the bound DNA into fragments.
- One strand is destroyed and the other penetrates the recipient bacterium.
- The Rec A protein promotes genetic exchange (recombination) between a fragment of the donor's DNA and the recipient's DNA.

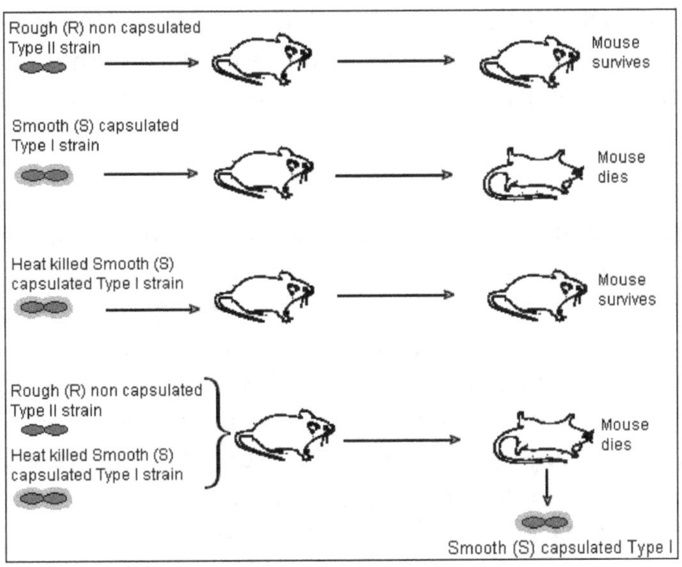

Some bacteria are able to take up DNA naturally. However, these bacteria only take up DNA a particular time in their growth cycle (log phase) when they produce a specific protein called a competence factor. Uptake of DNA by Gram positive and Gram negative bacteria differs. In Gram positive bacteria the DNA is taken up as a single stranded molecule and the complementary strand is made in the recipient. In contrast, Gram negative bacteria take up double stranded DNA.

Significance

Transformation occurs in nature and it can lead to increased virulence. In addition transformation is widely used in recombinant DNA technology.

Conjugation

In 1946 Joshua Lederberg and Tatum discovered that some bacteria can transfer genetic information to other bacteria through a process known as conjugation. Bacterial conjugation is the transfer of DNA from a living donor bacterium to a recipient bacterium.

Plasmids are small autonomously replicating circular pieces of double-stranded circular DNA. Conjugation involves the transfer of plasmids from donor bacterium to recipient bacterium. Plasmid transfer in Gram-negative bacteria occurs only between strains of the same species or closely related species. Some plasmids are designated as F factor (F plasmid, fertility factor or sex factor) because they carry genes that mediate their own transfer. The F factor can replicate autonomously in the cell. These genes code for the production of the sex pilus and enzymes necessary for conjugation. Cells possessing F plasmids are F^+ (male) and act as donors. Those cells lacking this plasmid are F^- (female) and act as recipient. All those plasmids, which confer on their host cells to act as donors in conjugation are called transfer factor.

Each Gram negative F^+ bacterium has 1 to 3 sex pili that bind to a specific outer membrane protein on recipient bacteria to initiate mating. The sex pilus then retracts, bringing the two bacteria in contact and the two cells become bound together at a point of direct envelope-to-envelope contact. In Gram-positive bacteria sticky surface molecules are produced which bring the two bacteria into contact. Gram-positive donor bacteria produce adhesins that cause them to aggregate with recipient cells, but sex pili are not involved. DNA is then transferred from the donor to the recipient. Plasmid-mediated conjugation occurs in Bacillus subtilis, Streptococcus lactis, and Enterococcus faecalis but is not found as commonly in the Gram-positive bacteria as compared to the Gram-negative bacteria.

F^+ Conjugation

This results in the transfer of an F^+ plasmid (coding only for a sex pilus) but not chromosomal DNA from a male donor bacterium to a female recipient bacterium. The two strands of the plasmid separate. One strand enters the recipient bacterium progressing in the 5' to 3' direction while one strand remains in the donor. The complementary strands are synthesized in both donor and recipient cells. The recipient then becomes an F+ male and can make a sex pilus. During conjugation, no cytoplasm or cell material except DNA passes from donor to recipient. The mating pairs can be separated by shear forces and conjugation can be interrupted. Consequently, the mating pairs remain associated for only a short time. After conjugation, the cells break apart. Following successful conjugation the recipient becomes F^+ and the donor remains F^+.

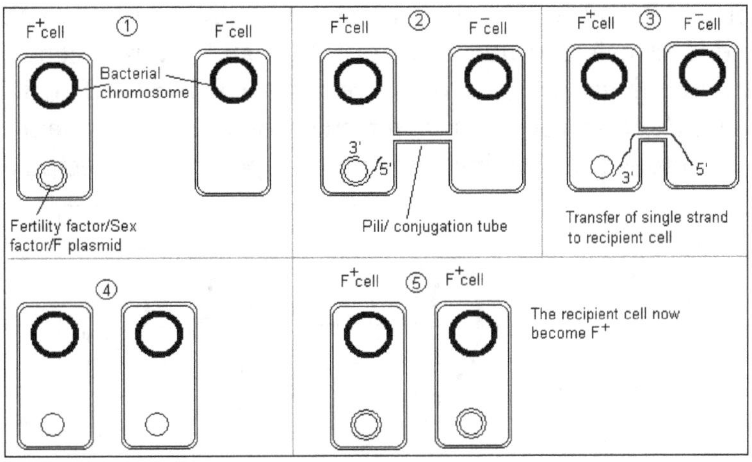

Resistance Plasmid Conjugation

Some Gram-negative bacteria harbor plasmids that contain antibiotic resistance genes; such plasmids are called R factors. The R factor has two components, one that codes for self-transfer (like F factor) called RTF (resistance transfer factor) and the other R determinant that contains genes coding for antibiotic resistance. R plasmids may confer resistance to as many as five different antibiotics at once upon the cell and by conjugation; they can be rapidly disseminated through the bacterial population. The difference between F factor and R factor is that the latter has additional genes coding for drug resistance. During conjugation there is transfer of resistance plasmid (Rplasmid) from a donor bacterium to a recipient. One plasmid strand enters the recipient bacterium while one strand remains in the donor. Each strand then makes a complimentary copy. R-plasmid has genes coding for multiple antibiotic resistance as well as sex pilus formation. The recipient becomes multiple antibiotic resistant and male, and is now able to transfer R-plasmids to other bacteria. When the recipient cells acquire entire R factor, it too expresses antibiotic resistance. Sometimes RTF may disassociate from the R determinant and the two components may exist as separate entities. In such cases though the host cell remains resistant to antibiotics, it cannot transfer this resistance to other cells. Sometimes RTF can have other genes (such as those coding for hemolysin, enterotoxin) apart from R determinants attached to it.

HFR (High Frequency Recombinant) Conjugation

Plasmids may integrate into the bacterial chromosome by a recombination event depending upon the extent of DNA homology between the two. After integration, both plasmid and chromosome will replicate as a single unit. A plasmid that is capable of integrating into the chromosome is called an episome. If the F plasmid is integrated into the chromosome it is called an Hfr cell. After integration, both chromosome and plasmid can be conjugally transferred to a recipient cell. Hfr cells are called so because they are able to transfer chromosomal genes to recipient cells with high frequency.

The DNA is nicked at the origin of transfer and is replicated. One DNA strand begins to passes through a cytoplasmic bridge to the F^- cell, where its complementary strand is synthesized. Along with the portion of integrated plasmid, the chromosome is also transmitted to the F^- cell. The bacterial connection usually breaks before the transfer of the entire chromosome is completed so the remainder of the F^+ plasmid rarely enters the recipient. Usually only a part of the Hfr chromosome as well as the plasmid is transferred during conjugation and the recipient cell does not receive complete F factor. After conjugation the Hfr cell remains Hfr but the F^- cell does not become F^+ and continues to remain F^-. However the transferred chromosome fragment recombines with the chromosome of F- cell thereby transferring some new property to the recipient cell.

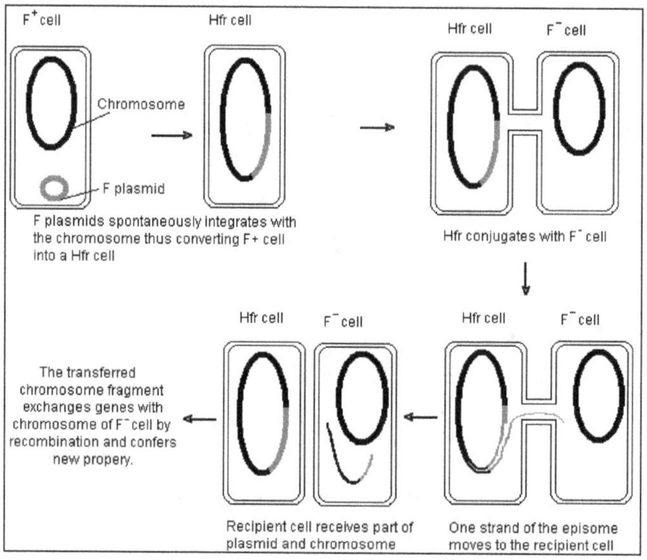

The integration of episome into the chromosome is not stable and the episomes are known to revert back to free state. While doing so, the episomes sometimes carry fragments of chromosomal genes along with it. Such an F factor that incorporates some chromosomal genes is called F prime (F') factor. When such a F' cell mates with F- recipient cell, it not only transfers the F factor but also the host genes that it carried with it. This process of transfer of chromosomal genes along with F factor is known is sexduction.

Significance

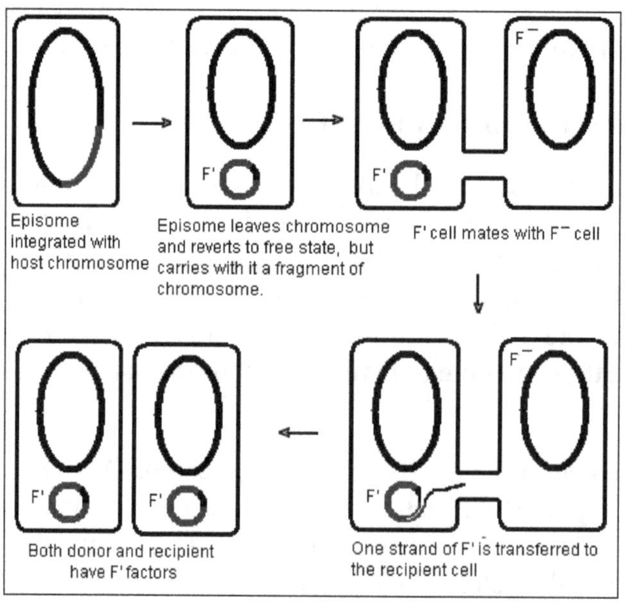

Among the Gram negative bacteria this is the major way that bacterial genes are transferred. Transfer can occur between different species of bacteria. Transfer of multiple antibiotic resistances by conjugation has become a major problem in the treatment of certain bacterial diseases. Since the recipient cell becomes a donor after transfer of a plasmid, an antibiotic resistance gene carried on a plasmid can quickly convert a sensitive population of cells to a resistant one.

Transduction

Bacteriophages are viruses that parasitize bacteria and use their machinery for their own replication. During the process of replication inside the host bacteria the bacterial chromosome or plasmid is erroneously packaged into the bacteriophage capsid. Thus newer progeny of phages may contain fragments of host chromosome along with their own DNA or entirely host chromosome. When such phage infects another bacterium, the bacterial chromosome in the phage also gets transferred to the new bacterium. This fragment may undergo recombination with the host chromosome and confer new property to the bacterium.

Life cycle of bacteriophage may either by lytic or lysogenic. In the former, the parasitized bacterial cell is killed with the release of mature phages while in the latter the phage DNA gets incorporated into the bacterial chromosome as prophage.

Following are the stages of transduction involving a lytic phage:

- A lytic bacteriophage adsorbs to a susceptible bacterium.
- The bacteriophage genome enters the bacterium. The phage DNA directs the bacterium's metabolic machinery to manufacture bacteriophage components and enzymes.
- Occasionally during maturation, a bacteriophage capsid incorporates a fragment of donor bacterium's chromosome or a plasmid instead of a phage genome by mistake.
- The bacteriophages are released with the lysis of bacterium.
- The bacteriophage carrying the donor bacterium's DNA adsorbs to another recipient bacterium.
- The bacteriophage inserts the donor bacterium's DNA it is carrying into the recipient bacterium.
- The donor bacterium's DNA is exchanged by recombination for some of the recipient's DNA.

In case of temperate phages that undergo lysogenic cycle, the phage DNA gets incorporated into the bacterium chromosome. This is called a prophage and it behaves as if it were a part of bacterial chromosome. This process is known as lysogenic conversion and the bacteria are called lysogenic bacteria. The genes present in the phage DNA also get expressed in the bacterium. Only those strains of Corynebacterium diphtheriae that have been lysogenised with beta prophage produce the diphtheria toxin. The prophage sometimes disassociates itself from the host chromosome during multiplication of lysogenic bacteria, and in doing so; it sometimes carries along with itself fragments of bacterial chromosome. The separated prophage then initiates lytic cycle and the subsequent phage progeny may have a piece of chromosomal DNA. When such phage infects another bacterium, newer characteristics coded by that chromosomal gene are conferred. Two types of transduction are known; restricted transduction and generalized transduction. Generalized transduction can transfer any bacterial gene to the recipient. This process may occur with phages (lytic phages) that degrade their host DNA into pieces the size of viral genomes. If these pieces are erroneously packaged into phage particles, they can be delivered to another bacterium in the next phage infection cycle. Phages P22 of Salmonella typhimurium and P1 and μ of E. coli carry out generalized transduction. In restricted transduction only those chromosomal genes that lie adjacent to the prophage are transmitted. The lambda phage that infects E.coli always transfers gal+ gene (responsible for galactose fermentation). Specialized transduction is only effective in transducing a few special bacterial genes while generalized transduction can transduce any bacterial gene.

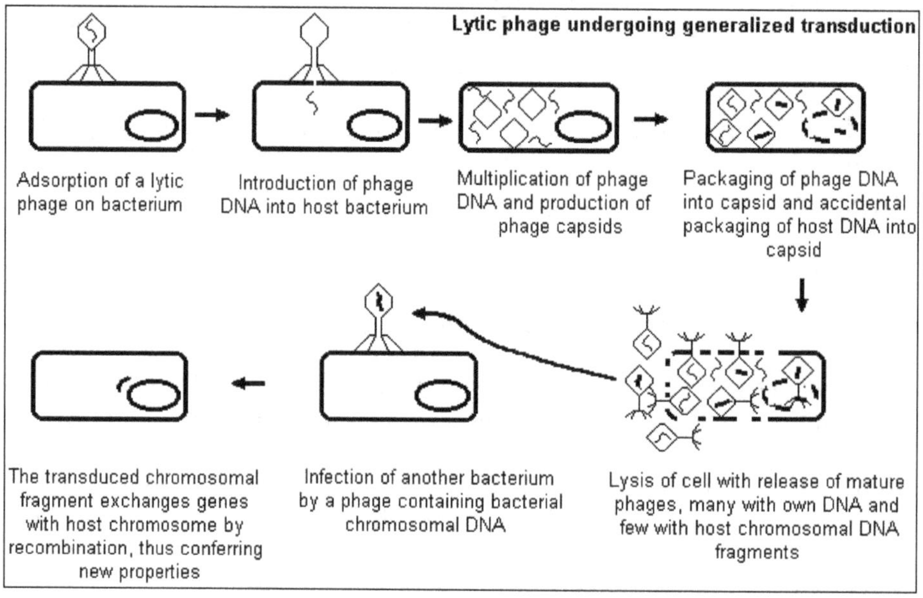

Plasmids

Plasmids are extra chromosomal elements found inside a bacterium. These are not essential for the survival of the bacterium but they confer certain extra advantages to the cell.

Number and size: A bacterium can have no plasmids at all or have many plasmids (20-30) or multiple copies of a plasmid. Usually they are closed circular molecules; however they occur as linear molecule in Borrelia burgdorferi. Their size can vary from 1 Kb to 400 Kb.

Multiplication: Plasmids multiply independently of the chromosome and are inherited regularly by the daughter cells.

Types of plasmids: R factor, Col factor, RTF and F factor.

F factor: This is also known as fertility factor or sex factor. Most plasmids are unable to mediate their own transfer to other cells. Vertical (inheritance) or horizontal (transfer) transmissions maintain plasmids. F factor is a plasmid that codes for sex pili and its transfer to other cells. Those bacteria that possess transfer factor are called F+, such bacteria have sex pili on their surface.

Those cells lacking this factor are designated F-. The F factor plasmid is transferred to other cells through conjugation. An F- cell will become F+ when it receives the fertility factor from another F+ cell.

R Factor

Those plasmids that code for the transmissible drug resistance are called R factor. These plasmids contain genes that code for resistance to many antibiotics. R factors may be transferred by conjugation and its transfer to other bacteria is independent of the F factor. Bacteria possessing such plasmids are resistant to many antibiotics and this drug resistance are transferred to closely related species. R factors may simultaneously confer resistance to five antibiotics. They are usually transferred to related species along with RTF.

Significance of Plasmids

- Codes for resistance to several antibiotics. Gram-negative bacteria carry plasmids that give resistance to antibiotics such as neomycin, kanamycin, streptomycin, chloramphenicol, tetracycline, penicillins and sulfonamides.
- Codes for the production of bacteriocines.
- Codes for the production of toxins (such as Enterotoxins by Escherichia coli, Vibrio cholerae, exfoliative toxin by Staphylococcus aureus and neurotoxin of Clostridium tetani).
- Codes for resistance to heavy metals (such as Hg, Ag, Cd, Pb etc.).
- Plasmids carry virulence determinant genes. E.g., the plasmid Col V of Escherichia coli contain genes for iron sequestering compounds.
- Codes resistance to UV light (DNA repair enzymes are coded in the plasmid).
- Codes for colonization factors that is necessary for their attachment. E.g., as produced by the plasmids of Yersinia enterocolitica, Shigella flexneri, Enteroinvasive Escherichia coli.
- Contains genes coding for enzymes that allow bacteria unique or unusual materials for carbon or energy sources. Some strains are used for clearing oil spillage.

Application of plasmids:

- Used in genetic engineering as vectors.
- Plasmid profiling is a useful genotyping method.

Episomes

Jacob and Wollman coined the term episome. Previously, it was considered synonymous with plasmids. F factors are those plasmids that can code for self-transfer to other bacteria. Occasionally such plasmids get spontaneously integrated into chromosome. Plasmids with this capability are called episomes and such bacterial cells are called Hfr cells i.e. high frequency of recombination.

Transposable Genetic Elements

Transposable genetic elements are segments of DNA that have the capacity to move from one location to another (i.e. jumping genes). Properties of Transposable Genetic Elements:

- Random movement: Transposable genetic elements can move from any DNA molecule to any DNA other molecule or even to another location on the same molecule. The movement is not totally random; there are preferred sites in a DNA molecule at which the transposable genetic element will insert.
- Not capable of self-replication: The transposable genetic elements do not exist autonomously and thus, to be replicated they must be a part of some other replicon.
- Transposition mediated by site-specific recombination: Transposition requires little or no homology between the current location and the new site. The transposition event is mediated by an enzyme transposase that is coded by the transposable genetic element. Recombination that does not require homology between the recombining molecules is called illegitimate or non-homologous recombination.
- Transposition can be accompanied by duplication: In many instances transposition of the transposable genetic element results in removal of the element from the original site and insertion at a new site. However, in some cases the transposition event is accompanied by the duplication of the transposable genetic element. One copy remains at the original site and the other is transposed to the new site.

Types of Transposable Genetic Elements

Insertion Sequences (IS)

Insertion sequences are transposable genetic elements that carry no known genes except those that are required for transposition. Insertion sequences are small stretches of DNA that have at their ends repeated sequences, which are involved in transposition. In between the terminal repeated sequences there are genes involved in transposition and sequences that can control the expression of the genes but no other nonessential genes are present.

Importance of IS

- Mutation: The introduction of an insertion sequence into a bacterial gene will result in the inactivation of the gene.
- The sites at which plasmids insert into the bacterial chromosome are at or near insertion sequence in the chromosome.
- Phase Variation: In Salmonella there are two genes, which code for two antigenically different flagellar antigens. The expression of these genes is regulated by an insertion sequences.

Transposons

Transposons are transposable genetic elements that carry one or more other genes in addition to those, which are essential for transposition. The structure of a transposon is similar to that of an insertion sequence. The extra genes are located between the terminal repeated sequences.

Importance of Transposons

Many antibiotic resistance genes are located on transposons. Since transposons can jump from one DNA molecule to another, these antibiotic resistance transposons are a major factor in the development of plasmids, which can confer multiple drug resistance on a bacterium harboring such a plasmid. These multiple drug resistance plasmids have become a major medical problem.

Genetic Mechanism of Drug Resistance

Antibiotic resistance in bacteria may either be intrinsic or acquired. Intrinsic resistance means that the bacteria were resistant to the antibiotic even before the antibiotic was introduced. Acquired resistance means that a bacterium that was previously sensitive to an antibiotic has now turned resistant. It is the acquired resistance that is of great importance because it would result in treatment failure as well as potential dissemination of resistance to other bacteria.

The physiological mechanisms of antibiotic resistance include:

- Inactivation of the antibiotic by enzymes produced by the bacteria.
- Alteration of target proteins such that the antibiotic doesn't bind or binds with decreased affinity.
- Alteration of the membrane which decreases the permeability of the antibiotic.
- Active efflux of the antibiotic.
- Development of alternate metabolic pathway to bypass the action of antibiotic.

The Genetic Mechanisms of Antibiotic Resistance Include

Mutations

Mutations can alter the protein to which the antibiotic must bind resulting in a protein with little or no affinity for the drug. Mutations can be either step-wise, as seen with Penicillin, where high levels of resistance are achieved by a series of small-step mutations. Multiple-drug resistance in Mycobacteria is apparently the result of the step-wise accumulation of resistance to individual drugs. Mutations can also be one-step where single mutation is sufficient to bring about resistance in the bacteria as in Streptomycin resistance in Mycobacterium tuberculosis. In case of tuberculosis, Mycobacteria are known to mutate during the course of treatment. Initially the antitubercular drug kills the bacteria but soon resistant mutants develop which would eventually replace the sensitive ones resulting in treatment failure. It is for this reason that multiple drugs are included in the treatment of tuberculosis.

Examples:

- Methicillin resistance in S.aureus due to mutation in Penicillin Binding Protein.

- Mutations in genes are associated with isoniazid and rifampin resistance in Mycobacterium tuberculosis.

Transferable Drug Resistance

Drug resistance can be transferred between related bacteria or different taxonomic groups by the process of conjugation and transduction. Drug resistance mediated by R plasmids is the most important method of drug resistance. Acquisition of R factor confers resistance to several antibiotics. Examples:

- Resistance to penicillins in Gram negative bacteria due to beta-lantanas enzymes coded by plasmids.
- Resistance to penicillin in Staphylococci due to beta-lactamase transferred by transduction.
- Resistance to Chloramphenicol in Salmonella due to chloramphenicol acetyl transferase coded by a plasmid.

Fungal Genetics

Fungi possess strikingly different morphologies. They include large, fleshy, and often colorful mushrooms or toadstools, filamentous organisms only just visible to the naked eye, and single-celled organisms such as yeasts. Molds are important agents of decay. They also produce a large number of industrially important compounds like antibiotics (penicillin , griseofulvin, etc.), organic acids (citric acid, gluconic acid, etc.), enzymes (alpha-amylases, lipase, etc.), traditional foods (softening and flavoring of cheese, shoyu soy sauce, etc.), and a number of other miscellaneous products (gibberellins, ergot alkaloids, steroid bioconversions). As late as 1974 the only widely applicable techniques for strain improvement were mutation, screening, and selection. While these techniques proved dramatically successful in improving penicillin production, they deflected attempts to employ a more sophisticated approach to genetic manipulation. The study of fungal genetics has recently changed beyond all recognition.

The natural genetic variation present in fungal species has been characterized using molecular methods such as electrophoretic karyotyping, restriction fragment length polymorphism, DNA finger printing, and DNA sequence comparisons. The causes for the variation include chromosomal polymorphism, changes in repetitive DNA, transposons, virus-like elements, and mitochondrial plasmids.

Genetic recombination occurs naturally in many fungi. Many industrially important fungi such as Aspergilli and Penicillia lack sexuality, so in these species parasexual systems (cycles) provide the basis for genetic study and breeding programs. The parasexual cycle is a series of events that can be induced when two genetically different strains are grown together in the laboratory. A heterokaryon, which is mycelium with two different nuclei derived from two different haploid strains, is produced by the fusion of hyphae. Increased penicillin titer in the haploid progeny of parasexual crosses has been achieved in Penicillium chrysogenum. A more direct approach has been developed using protoplasts. These are isolated from vegetative cells of fungi or yeasts by removing the cell wall by digestion using a cell wall degrading enzyme. Protoplasts from the two strains can be fused by treatment with polyethylene glycol. Protoplast fusion in fungi initiates the parasexual cycle, resulting in the formation of diploidy and mitotic recombination and segregation. A selection procedure to screen such fusants is done using genetic markers. A good example of applying this technique is the fusion of a fast growing but poor glucoamylase producer with a slow growing but excellent producer of glucoamylase. The desired result will be a strain that is both fast growing and an excellent producer of enzyme.

The realization that transformation of genetic material into fungi can occur came with the discovery that yeasts like Saccharomyces cerevisiae and filamentous fungi like Podospora anserine contain plasmids. Currently transformation technology is largely based on the use of Neurospora crassa and Aspergillus nidulans, though methods for use in filamentous organisms are being developed. The protocols used in transformation of filamentous fungi involve cloning the desired gene into the plasmid from E. coli or a plasmid constructed from genetic material from E. coli and Saccharomyces cerevisiae. Protoplasts from the recipient strains are then formed and mixed with the plasmid. After incubating for a short time to allow for the uptake of the plasmid DNA, the protoplasts are allowed to regenerate and the cells are screened for the presence of the specific marker.

The application of recombinant DNA to yeasts and filamentous fungi has opened up new possibilities in relation to the construction of highly productive strains. The filamentous fungi are now established as potent host organisms for the production of heterologous proteins. This is particularly useful as expression of specific proteins can reach relatively high levels. Using Aspergillus as a host for reproduction has led to the production of many recombinant products like human therapeutic proteins, including growth factors, cytokines , and hormones. While expression can be good in E. coli, lack of posttranslational modifications has limited their usage. The use of Saccharomyces species has not been highly successful for the production of extracellular proteins. Most of the initial advances for the production of heterologous proteins have been with filamentous fungi, namely Aspergillus nidulans. Although this organism is not of industrial importance it is nevertheless genetically well characterized; in addition, this organism has secretion signals that result in recombinant proteins being identical to mammalian cells. This allows the product from such systems to be used safely in human therapy. Other systems that have been used include Pichia and Trichoderma, which have been widely used in industry. Now that the complete genome of S. cerevisiae has been deciphered, and with more fungi genomes in the pipeline, an even better understanding of fungal genetics is certain.

Viral Genetics

Viruses are simple entities, lacking an energy-generating system and having very limited biosynthetic capabilities. The smallest viruses have only a few genes; the largest viruses have as many as 200. Genetically, however, viruses have many features in common with cells. Viruses are subject to mutations, the genomes of different viruses can recombine to form novel progeny, the expression of the viral genome can be regulated, and viral gene products can interact. By studying viruses, we can learn more about the mechanisms by which viruses and their host cells function.

Genetic Change in Viruses

Two principal mechanisms are involved: mutation and recombination. Alterations in the genetic material of a virus may lead to changes in the function of viral proteins. Such changes may result in the creation of new viral serotypes or viruses of altered virulence.

Mutations

Mutations arise by one of three mechanisms:

- By the effects of physical mutagens (UV light, x-rays) on nucleic acids.
- By the natural behavior of the bases that make up nucleic acids (resonance from keto to enol and from amino to imino forms).
- Through the fallibility of the enzymes that replicate the nucleic acids.

The first two mechanisms act similarly in all viruses; hence, the effects of physical mutagens and the natural behavior of nucleotides are relatively constant. However, viruses differ markedly in their mutation rates, which is due primarily to differences in the fidelity with which their enzymes replicate their nucleic acids. Viruses with high-fidelity transcriptases have relatively low mutation rates and vice versa.

Mutation Rates and Outcomes

DNA viruses have mutation rates similar to those of eukaryotic cells because, like eukaryotic DNA polymerases, their replicatory enzymes have proofreading functions. The error rate for DNA viruses has been calculated to be 10^{-8} to 10^{-11} errors per incorporated nucleotide. With this low mutation rate, replication of even the most complex DNA viruses, which have 2×10^5 to 3×10^5 nucleotide pairs per genome, will generate mutants rather rarely, perhaps once in several hundred to many thousand genome copies. The RNA viruses, however, lack a proofreading function in their replicatory enzymes, and some have mutation rates that are many orders of magnitude higher—10^{-3} to 10^{-4} errors per incorporated nucleotide. Even the simplest RNA viruses, which have about 7,400 nucleotides per genome, will generate mutants frequently, perhaps as often as once per genome copy.

Not all mutations that occur persist in the virus population. Mutations that interfere with the essential functions of attachment, penetration, un-coating, replication, assembly, and release do not permit mis-replication and are rapidly lost from the population. However, because of the redundancy of the genetic code, many mutations are neutral, resulting either in no change in the viral protein or in replacement of an amino acid by a functionally similar amino acid. Only mutations that do not cripple essential viral functions can persist or become fixed in a virus population.

Phenotypic Variation by Mutations

Mutations that alter the viral phenotype but are not deleterious may be important. For example, mutation can create novel antigenic determinants. A mutation in the hemagglutinin gene of influenza A virus can give rise to a hemagglutinin molecule with an altered antigenic site (epitope). Provided the attachment function of the new hemagglutinin is intact, the mutant virus may be able to initiate an infection in an individual immune to viruses expressing the previous hemagglutinin.

For example, from 1968 to 1979, mutations altered 10 percent of the amino acids in the influenza virus hemagglutinin serotype H3 molecule. This relatively modest mechanism of antigenic change through mutation, called antigenic drift, may allow a virus to outflank host defenses and cause disease in previously immune individuals.

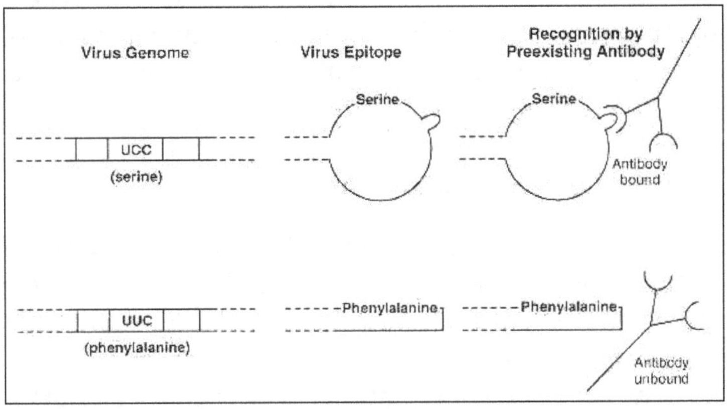

Figure: Mutation causing phenotypic (antigenic) variation.

Mutation of the codon for the hydrophilic amino acid serine to the codon for the hydrophobic amino acid phenylalanine can change an epitope on the viral hemagglutinin protein and thereby alter its recognition by specific antibody. The mutant virus may then be able to infect a previously immune host.

Vaccine Strains from Mutations

Mutation has been a principal tool of virologists in developing attenuated live virus vaccines. For example, the Sabin vaccine strains of poliovirus were developed by growing polioviruses in monkey kidney cells. Mutation and selection produced variant polioviruses that were adapted for efficient replication in these cells. Some of the mutations in these variants affected the genes coding for the poliovirus coat proteins in such a way as to produce mutants unable to attach to human neural cells but still able to infect human intestinal cells. Infection of human intestinal cells does not produce paralytic disease but does induce immunity. Poliovirus vaccine strains 1 and 2 have multiple mutations in the coat proteins and are very stable. The type 3 vaccine strain is less stable and is subject to back-mutations (reversions) that restore neural virulence. This vaccine strain therefore causes paralytic disease in one out of every several million vaccinated individuals. Despite the possibility of back-mutations, the generation and selection of attenuated viral mutants remains an important mechanism for producing viral vaccines.

Table: Live Attenuated Virus vaccines.

Viral Vaccine	Cell used for Production of Attenuated virus
Measles virus (rubeola virus)	Chick embryo cells
German measles virus (rubella virus)	Monkey Kidney cells
Mumps virus	Embryonated chicken eggs
Poliovirus	Monkey kidney cells
Yellow fever virus	Chick embryo cells
Varicella-Zosta virus	Human embryo cells

Recombination

Viral recombination occurs when viruses of two different parent strains coinfect the same host cell and interact during replication to generate virus progeny that have some genes from both parents. Recombination generally occurs between members of the same virus type (e.g., between two influenza viruses or between two herpes simplex viruses). Two mechanisms of recombination have been observed for viruses: independent assortment and incomplete linkage. Either mechanism can produce new viral serotypes or viruses with altered virulence.

Recombination by Independent Assortment

Independent assortment occurs when viruses that have multipartite (segmented) genomes trade segments during replication. These genes are unlinked and assort at random. Recombination by independent assortment has been reported, for example, for the influenza viruses and other orthomyxoviruses (8 segments of single-stranded RNA) and for the reoviruses (10 segments of double-stranded RNA). The frequency of recombination by independent assortment is 6 to 20 percent for orthomyxoviruses. Independent assortment between an animal and a human strain of influenza virus during a mixed infection can yield an antigenically novel influenza virus strain capable of infecting humans but carrying animal-strain hemagglutinin and/or neuraminidase surface molecules. This recombinant can infect individuals that are immune to the parent human virus. This mechanism results in an immediate, major antigenic change and is called antigenic shift. Antigenic shifts in influenza virus antigens can give rise to pandemics (worldwide epidemics) of influenza. Such antigenic shifts have occurred relatively frequently during recent history. Because the number of different serotypes of hemagglutinin and neuraminidase are limited, a given strain reappears from time to time. For example, the H1N1 influenza virus strain was responsible for the 1918 to 1919 influenza pandemic that caused 20 million deaths. The same virus also caused pandemics in 1934 and in 1947, then disappeared after 1958 and reappeared in 1977. The reappearance of virus strains after an absence is believed to be the result of recombinational events involving the independent assortment of genes from two variant viruses.

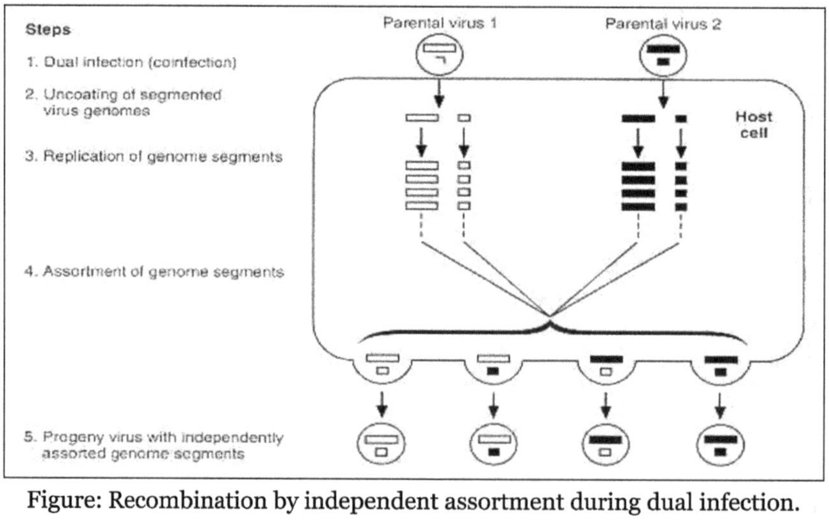

Figure: Recombination by independent assortment during dual infection.

After infection of a cell with two viruses with two or more genetic segments ("chromosomes"), reassortment of the replicated segments can occur. Independent assortment results in the generation of progeny viruses whose genomes contain segments of genome from both types of parental viruses.

Table: Antigenic shifts resulting from Reassortment of Genome segments.

Year	Strain
1890	H2N8
1990	H3N8
1918	H1N1
1957	H2N2 (Asian Flu)
1968	H3N2 (Hong Kong Flu)
1977	H1N1

Recombination of Incompletely Linked Genes

Recombination also occurs between genes residing on the same piece of nucleic acid. Genes that generally segregate together are called linked genes. If recombination occurs between them, the linkage is said to be incomplete. Recombination of incompletely linked genes occurs in all DNA viruses that have been studied and in several RNA viruses.

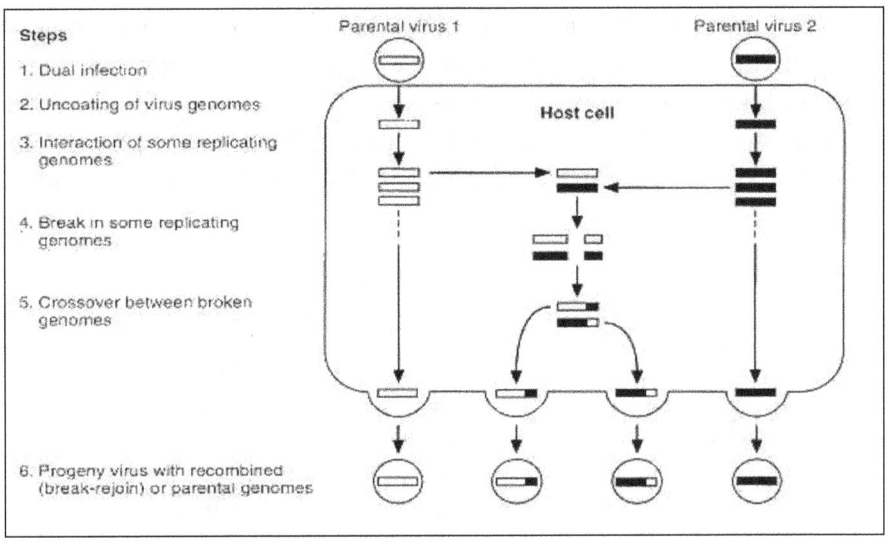

Figure: Recombination by break-rejoin of incompletely linked genes.

The genetic interaction of DNA viruses can result in break-rejoin recombination, in which the two DNA molecules of different viruses break and then cross over. Break-rejoin recombination results in novel progeny viruses with some DNA sequences of both types of parental viruses.

In DNA viruses, as in prokaryotic and eukaryotic cells, recombination between incompletely linked genes occurs by means of a break-rejoin mechanism. This mechanism involves the actual severing of the covalent bonds linking the bases of each of the two DNA strands in a DNA molecule. The severed DNA strands are then rejoined to the DNA strands of a different DNA molecule that has been broken in a similar site. Recombination rates for herpesviruses, which are DNA viruses that replicate in the nucleus of infected cells, approximate those expected for a eukaryotic genome of the size of the herpesvirus genome. Herpesviruses have an average recombination frequency of 10 to 20 percent for any two loci. However, the rate of recombination between a specific pair of genetic loci depends on the distance between them and varies from less than 1 percent to approximately 50 percent. Measurement of the recombination frequencies for different loci can be used to map the virus genome. In this type of genetic map, loci with high recombination frequencies are far apart and loci with low recombination frequencies are close together.

Recombination has been shown to occur in several positive-sense single-stranded RNA virus groups: retroviruses, picornaviruses, and coronaviruses. That is initially surprising, as recombination between RNA molecules has not been observed in prokaryotic or eukaryotic cells. In retroviruses, recombination actually occurs at the point in replication when the retrovirus genome is in a DNA form and takes place by the same break-rejoin mechanism as in cells and DNA viruses. Recombination can occur both between two related retroviruses and between the retrovirus DNA and the host cell DNA. Recombination between two retroviruses gives rise to novel viral progeny with reassorted genes. Recombination between retroviruses and the host cell can give rise to novel viral progeny that carry non-viral genes. If these host genes code for growth factors, growth factor receptors, or a number of other specific cellular proteins, the recombinant retroviruses may be oncogenic.

In picornaviruses and coronaviruses, recombination takes place at the level of the interaction of the viral RNA genomes and is not believed to occur by a break-rejoin mechanism. The mechanism is currently believed to be a copy-choice mechanism. Copy-choice may occur in these RNA viruses because the viral RNA polymerase binds to only a few bases of the template RNA at any one time. Such a weak interaction of the polymerase with the template RNA would permit the polymerase, carrying its RNA strand, to disassociate from the original template nucleic acid strand and then associate with a new template RNA strand. Recombination frequencies in the range of 0.2 to 0.4 percent have been reported. Therefore, the efficiency of this mechanism of recombination is low.

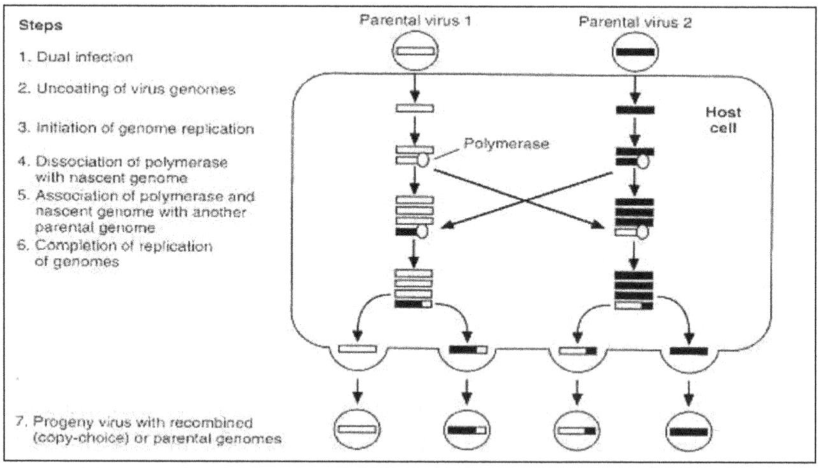

Figure: Recombination by copy-choice of incompletely linked genes. The genetic interaction of certain RNA viruses can result in copy-choice recombination.

In this mechanism, the polymerase begins replicating RNA template. By an unknown mechanism, which may involve a high degree of secondary structure in the viral RNA template, the polymerase complex (with its nascent viral RNA molecule) reassociates with the template viral RNA of different parental virus, such that the nascent RNA molecule will be synthesized as a novel recombinant virus whose RNA genome contains genes from each parental virus type.

Phenotypic Variation from Recombination

As mentioned above, viral recombination is important because it can generate novel progeny viruses that express new antigenic and/or virulence characteristics. For example, the novel progeny viruses may have new surface proteins that permit them to infect previously resistant individuals; they may have altered virulence characteristics; they may have novel combinations of proteins that make them infective to new cells in the original host or to new hosts; or they may carry material of cellular origin that gives them oncogenic potential.

Vaccines and Gene Therapy through Recombination

Recombination is being used experimentally by virologists to create new vaccines. Vaccinia virus, a DNA virus of the poxvirus group, was used as a live vaccine in the eradication of smallpox. Recombinant vaccinia viruses are being developed that carry vaccinia virus DNA recombined with DNA from other sources (exogenous DNA). For example, vaccinia virus strains carrying DNA coding for bacterial and viral antigens have been produced. It is expected that after vaccination with the recombinant vaccinia virus, the bacterial or viral antigen (immunogen) will be produced. The presence of this immunogen will then stimulate specific antibody production by the host, resulting in protection of the host from the immunogen. Studies with these live, recombinant vaccinia viruses are currently under way to determine whether inoculation of the skin with the recombinant virus can induce a protective host antibody response to the bacterial or viral antigens. Other studies are investigating the use of live, recombinant adenoviruses containing bacterial or viral genes to infect the gastrointestinal tract and induce both mucosal and systemic immunity.

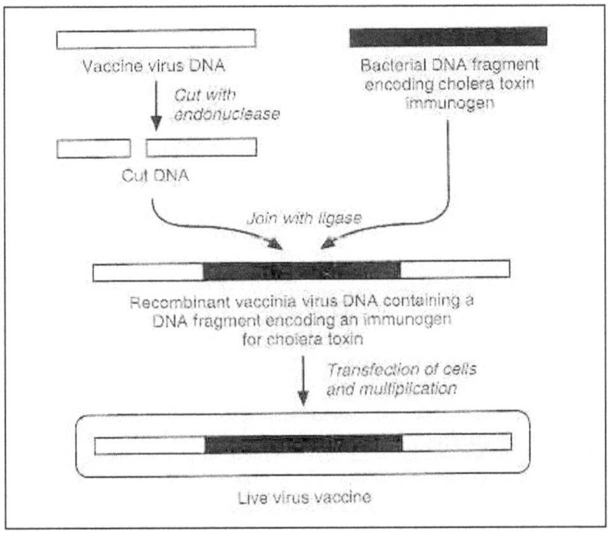

Figure: Development of recombinant vaccinia virus for immunization against cholera toxin.

Vaccinia virus genomic DNA is cut with an endonuclease. A specific sequence of DNA (with appropriate regulatory sequences) coding for a protein (e.g., cholera toxin) to be used as an immunogen is ligated into the vaccinia virus genome, making a recombinant vaccinia virus. The DNA will be transcribed and the immunogen will be produced along with vaccinia virus proteins in infected cells following vaccination. The immunogen will then elicit antibody production by the host, providing protective immunity.

In a similar manner, recombinant viruses are also being developed that carry normal human genes. It is envisioned that such recombinant viruses could be useful for gene therapy. Target diseases for gene therapy span a wide range, including diabetes, cystic fibrosis, severe combined immunodeficiency syndrome, etc.

References

- Introduction-to-microbial-genetics, microbial-genetics, microbiology, biology: cliffsnotes.com, Retrieved 13, August 2020
- Genetics, micronotes: microrao.com, Retrieved 05, May 2020

Chapter 4

Nutrition and Growth

All microbes like other living organisms need food for survival and growth. Their food may contain various forms of nutrients like carbon, oxygen, hydrogen, phosphorus and sulfur. An increase of the microbial population, subject to favorable conditions, is known as microbial growth. Such growth may be induced in controlled scientific environments through culture media. This chapter discusses microbial nutrition and growth for a thorough understanding of the subject.

Microbial Nutrient Requirements

Microbial cell composition shows that 95% of cell dry weight is made up of a few major elements: Carbon, oxygen, hydrogen, nitrogen, sulfur, phosphorous, potassium, calcium, magnesium and iron.

Macronutrients or macro elements:

These are required by microorganisms in relatively large amounts. Carbon, oxygen, hydrogen nitrogen, sulfurs and phosphorous are components of carbohydrates, lipids, proteins and nucleic acids. The remaining four macro elements (K, Ca, Mg and Fe) exist in the cell as cations.

K^+ - is required for the activity by a number of enzymes, including those involved in protein synthesis.

Ca^{2+} - contributes to the heat resistance of bacterial endospores. 15% of spore contains dipicolinic acid and calcium.

Mg^{2+} - serves as a cofactor for many enzymes, complexes with ATP and stabilizes ribosomes and cell membranes.

Fe^{2+} and Fe^{2+} - part of cytochromes and a cofactor for enzymes and electron-carrying proteins.

Micronutrients or Trace elements:

These are manganese, zinc, cobalt, molybdenum, nickel and copper. These are normally part of enzymes and cofactors, and they aid in the catalysis of reactions and maintenance of protein structure.

Zn^{2+} - is present at the active site of some enzymes but is also involved in the association of regulatory and catalytic subunits in *E. coli* aspartate carbomoyl transferase.

Mn^{2+} - aids many enzymes catalyzing the transfer of phosphate groups.

Mo^{2+} - required for nitrogen fixation.

Co^{2+} - is a component of Vitamin B12.

Besides macro and micro nutrients, some microorganisms may have particular requirements that reflect the special nature of their morphology or environment. Diatoms need silicic acid to construct their beautiful cell walls of silica. Bacteria growing in saline lakes and oceans depend on the presence of high concentrations of sodium ion. Microorganisms require a balanced mixture of all the above nutrients for proper growth.

Requirements for carbon, hydrogen and oxygen:

Carbon is needed for the skeleton or backbone of all organic molecules and molecules serving as carbon sources normally also contribute both oxygen and hydrogen atoms. One important carbon source that does not supply hydrogen or energy is CO_2 . Autotrophs – can use CO_2 as their sole or principal source of carbon. Many microorganisms are autotrophic, and most of these carry out photosynthesis and use light as their energy source. Some autotrophs oxidize inorganic molecules and derive energy from electron transfer. Heterotrophs – are organisms that use reduced pre-formed organic molecules as carbon sources. Ex. Glycolytic pathway produces carbon skeleton for use in biosynthesis and also releases energy as ATP and NADH. Actinomycetes will degrade amyl alcohol, paraffin and even rubber. *Burkholderia cepacia* can use over 100 different carbon compounds. Some microorganisms can metabolize even relatively indigestible human-made substances such as pesticides. Indigestible molecules can be oxidized and degraded in the presence of a growth promoting nutrient that is metabolized at the same time, a process called Co-metabolism. The products of this breakdown can then be used as nutrients by other microorganisms.

Nutritional types of microorganisms:

In addition to Carbon, hydrogen and oxygen all organisms require sources of energy and electrons for growth.

Carbon sources:

Autotrophs - CO_2 sole or principal biosynthetic carbon source.

Heterotrophs – reduced, preformed organic molecules from other organisms.

Energy sources:

Phototrophs – use light as their energy source.

Chemotrophs – obtain energy from the oxidation of chemical compounds (either organic or in organic).

Electron sources:

Lithotrophs – use reduced inorganic substances as their electron source.

Organotrophs – extract electrons from organic compounds.

Four major nutritional classes based on their primary sources of carbon, energy and electrons is known.

Phtotolithotrophic autotrophs or photoautotrophs or photolithoautotrophs:

Source of energy – light energy.

Source of electrons – Inorganic hydrogen/ electron.

Carbon source - CO_2.

Example: Algae, purple and green sulfur bacteria and cyanobacteria.

Photoorganotrophic heterotrophy or photoorganoheterotrophy:

Source of energy – light energy.

Source of electrons – organic hydrogen/ electron.

Carbon source – organic carbon sources (CO_2 may also be used).

Example: Purple and green nonsulfur bacteria (common inhabitants of lakes and streams).

Chemolithotrophic autotrophs or chemolithoautotrophy:

Source of energy – Chemical energy source (inorganic).

Source of electrons – Inorganic hydrogen/ electron donor.

Carbon source - CO_2.

Example: Sulfur-oxidizing bacteria, hydrogen bacteria, nitrifying bacteria, iron-oxidizing bacteria.

Chemoorganotrophic heterotrophs or chemoorganoheterotrophy:

Source of energy – Chemical energy source (organic).

Source of electrons – Inorganic hydrogen/ electron donor.

Carbon source – organic carbon source.

Example: Protozoan, fungi, most non-photosynthetic bacteria (including most pathogens).

The most common nutritional types are photolithoautotrophs and chemoorganoheterotrophs. Bacteria *Beggiatoa* rely on inorganic energy sources and organic (or sometimes CO_2) carbon sources. These microbes are sometimes called Mixotrophic because they combine chemolithoautotrophic and heterotrophic metabolic processes.

Requirements for nitrogen, phosphorous and sulfur:

Nitrogen is needed for the synthesis of amino acids, purines, pyramidines, some carbohydrates and lipids, enzyme cofactors and other substances. Most phototrophs and many nonphotosynthetic microorganisms reduce nitrate to ammonia and incorporate the ammonia in assimilatory nitrate reduction. A variety of bacteria like many Cyanobacteria and *Rhizobiium* can reduce and assimilate atmospheric nitrogen using the nitrogenase systems. Phosphorous is present in nucleic acids, phospholipids, ATP, several cofactors, some proteins and other cell components. All microorganisms use inorganic phosphate as their phosphorous source and incorporate it directly. *E. coli* can use both organic and inorganic phosphate. Organophosphates such as hexose 6- phosphate can be taken up directly by transport proteins. Other organophosphates are often hydrolyzed in the periplasm by the enzyme alkaline phosphatase to produce inorganic phosphate which is then transported across the plasma membrane. When inorganic phosphate is outside the bacterium, it crosses the outer membrane by the use of a porin protein channel. Sulfur is needed for the synthesis of substances like the amino acids cysteine and methionine, some carbohydrates biotin and thiamine. Most of them use sulfate as a source of sulfur and reduce it by assimilatory sulfate reduction; a few require a reduced form of sulfur such as cysteine.

Growth factors:

Many microorganisms have the enzymes and pathways necessary to synthesize all cell components. Many lack one or more enzymes and hence require organic compounds because they are essential cell components or precursors of such components and cannot be synthesized by the organisms are called – growth factors. There are three major classes of growth factors:

Amino acids – needed for protein synthesis.

Purines and Pyramidines – for nucleic acid synthesis.

Vitamins – small organic molecules that usually make up all or part of enzyme cofactors, and only very small amounts sustain growth.

Knowledge of the specific growth factor requirements of many microorganisms makes possible quantitative growth response assays for a variety of substances. The observation that many microorganisms can synthesize large quantities of vitamins has led to their use in industry. Several water-soluble and fat-soluble vitamins are produced using industrial fermentations.

Ribofalvin – *Clostridium, Candida, Ashbya, Eremothecium.*

Coenzyme A – *Brevibacterium.*

Vitamin B_{12} – *Streptomyces, Propionibacterium, Pseudomonas.*

Vitamin C – *Gluconobacter, Erwinia, Corynebacterium.*

β- Carotene – *Dunaliella.*

Vitamin D - *Saccharomyces.*

Uptake of Nutrients

Uptake of the required nutrients by the microbial cell is important. Since microorganisms live in nutrient poor habitats, they must be able to transport nutrients from dilute solutions into the cell against concentration gradient. Finally, they must pass through a selectively permeable plasma membrane. Microorganisms use different transport mechanisms like facilitated diffusion, active transport and group translocation. Eukaryotic microorganisms do not employ group translocation but take up nutrients by endocytosis.

Movement of materials across the plasmamembrane is mostly done by two processes:

Passive processes : Substances cross the area from an area of high concentration to an area of low concentration without any expenditure of energy (ATP). Example, simple diffusion, osmosis and facilitated diffusion.

Active process: The cell must use energy (ATP) to move substances from areas of low concentration to areas of high concentration. Example, Group translocation.

Passive processes:

Passive or simple diffusion :Often called diffusion, is the process in which molecules move from a region of higher concentration to one of lower concentration. The rate is dependent on the size of the concentration gradient between a cell's exterior and its interior. Very small molecules such as water and oxygen and carbon dioxide move across membranes by simple or passive diffusion. Larger molecules, ions, and polar substances do not cross membranes by this method.

Osmosis: Is the net movement of solvent molecules across a selectively permeable membrane from an area in which the solvent molecules are highly concentrated to an area of low concentration until equilibrium is reached. In living systems the chief solvent is water. The three types of solutions which are normally found are isotonic, hypotonic and hypertonic.

Facilitated diffusion: The rate of diffusion across selectively permeable membrane is greatly increased by using carrier proteins, sometimes called permeases which are embedded in the plasma membrane. Because a carrier aids the diffusion process, it is called as facilitated diffusion. Carrier proteins also resemble enzymes in their specificity for the substances to be transported; each carrier is selective and will transport only closely related solutes. Because there is no energy input, molecules will continue to enter only as long as their concentration is greater on the outside. Two widespread major intrinsic protein channels in bacteria are aquaporins that transport water and glycerol facilitators which aid glycerol diffusion. The carrier protein complex spans the membrane. After the solute molecule binds to the outside, the carrier may change conformation and release the molecule on the cell interior. The carrier would subsequently change back to its original shape and be ready to pick up another molecule. The mechanism is driven by concentration gradients and therefore is reversible.

Examples. Glycerol is transported by facilitated diffusion in *E. coli, Salmonella typhimurium, Pseudomonas, Bacillus* and many other bacteria. This is prominent in eukaryotes where it is used to transport a variety of sugars and amino acids.

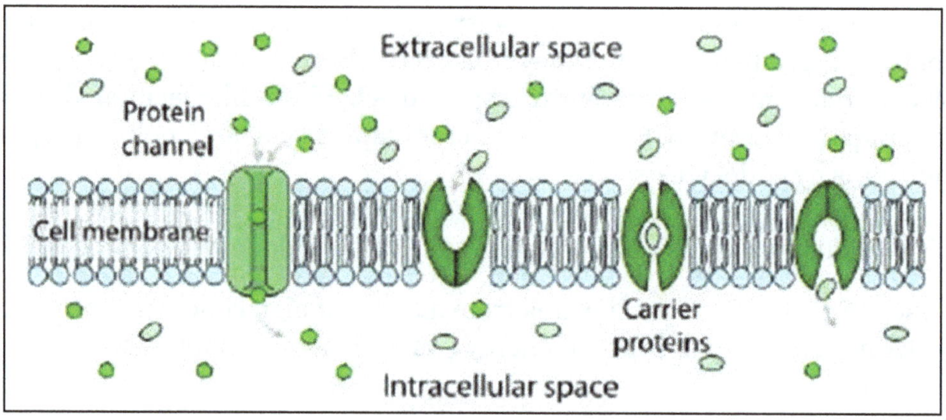

Facilitated diffusion. The carrier proteins aid in the release of solute molecules from extracellular space to the intracellular space.

Active Transport

Active transport is the transport of solute molecules to higher concentrations or against a concentration gradient, with the use of metabolic energy input. It resembles facilitated diffusion in the involvement of protein carrier activity, but differs in its use of metabolic energy and in its ability to concentrate substances. One example of active transport is binding protein transport systems or ATP-binding cassette transporters (ABC transporters) are active in bacteria, archaea and eukaryotes.

ABC transporters: These transporters consist of two hydrophobic membrane spanning domains associated on their cytoplasmic surfaces with two nucleotide-binding domains. The membrane spanning domains from a pore in the membrane and the nucleotide binding domains bind and hydrolyze ATP to drive uptake. ABC transporters employ special substrate binding proteins, which are located in the periplasmic space of gram-negative bacteria or are attached to membrane lipids on the external face of the gram positive plasma membrane. These binding proteins, bind the molecule to be transported and then interact with the membrane transport proteins to move the solute molecule inside the cell. *E. coli* transports a variety of sugars (arabinose, maltose, galactose, and ribose) and amino acids (glutamate, histidine, leucine) by this mechanism.

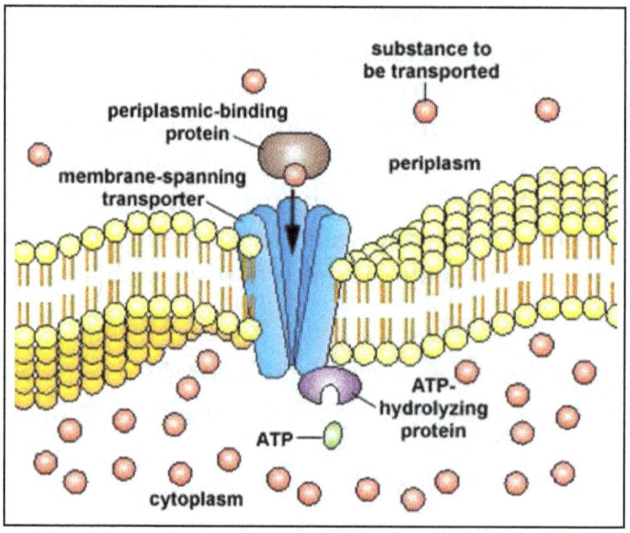

ABC transporter system.

Eukaryotic ABC transporters are sometimes of great medical importance. Some tumor cells pump drugs out using these transporters. Cystic fibrosis results from a mutation that inactivates an ABC transporter that acts as a chloride ion channel in lungs.

Bacteria also use Proton gradients generated during electron transport to drive active transport. The lactose permease of *E. coli* transports a lactose molecule inward as a proton simultaneously enters the cell. Such linked transport of two substances in the same direction is called Symport. *E. coli* also uses proton symport to take up amino acids and organic acids like succinate and malate.

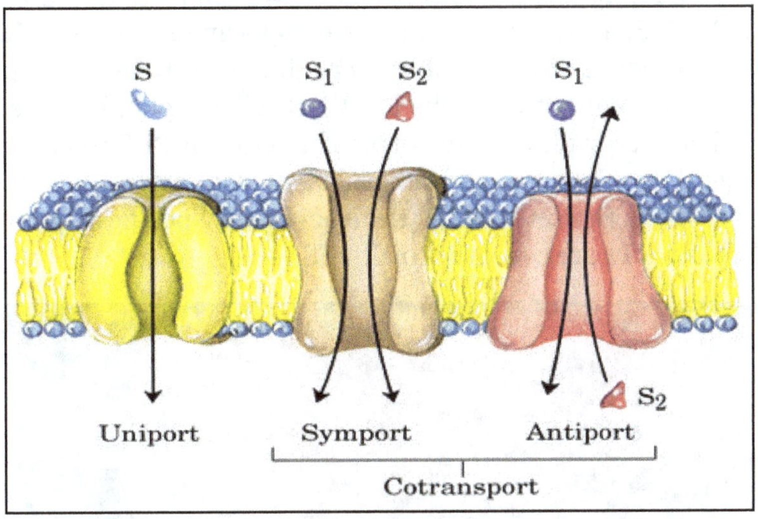

Comparison of the three different types of transport mechanisms.

A proton gradient also can power active transport indirectly, often through the formation of a sodium ion gradient. In *E. coli* , sodium transport system pumps sodium outward in response to the inward movement of protons. Such linked transport in which the transported substances move in opposite directions is called Anitport. The sodium gradient generated by this proton anitport system then drives the uptake of sugars and amino acids. *E. coli* has at least transport systems for the sugar galactose.

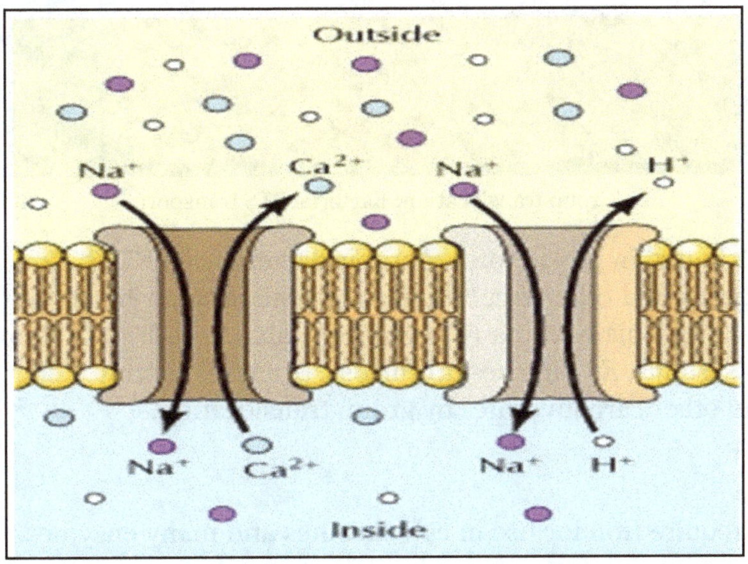

Antiport mechanism - linked transport of two substances in opposite direction.

Group Translocation: In active transport, solute molecules move across a membrane without modification. Many prokaryotes also take up molecules by group translocation, a process in which a molecule is transported into the cell while being chemically altered. For example, Phosphoenolpyruvate: Sugar phosphotransferease system (PTS). It transports a variety of sugars while phosphorylating them using phosphoenolpyruvate (PEP) as the phosphate donor.

PEP + Sugar (outside) → Pyruvate + Sugar-P (inside)

In *E. coli* and *Salmonella typhimurium*, it consists of two enzymes and a low molecular weight heat stable protein (HPr). HPr and enzyme I (EI) are cytoplasmic. Enzyme II (EII) is more variable in structure and often composed of three subunits or domains. EIIA is cytoplasmic and soluble. EIIB also is hydrophilic but frequently attached to EIIC, a hydrophobic protein that is embedded in the membrane. A high energy phosphate is transferred from PEP to enzyme II (EII) with the aid of enzyme I (EI) and HPr. Then a sugar molecule is phosphorylated as it is carried across the membrane by enzyme II (EII). Enzyme II (EII) transports only specific sugars and varies with PTS, whereas enzyme I (EI) and HPr are common to all PTS's.

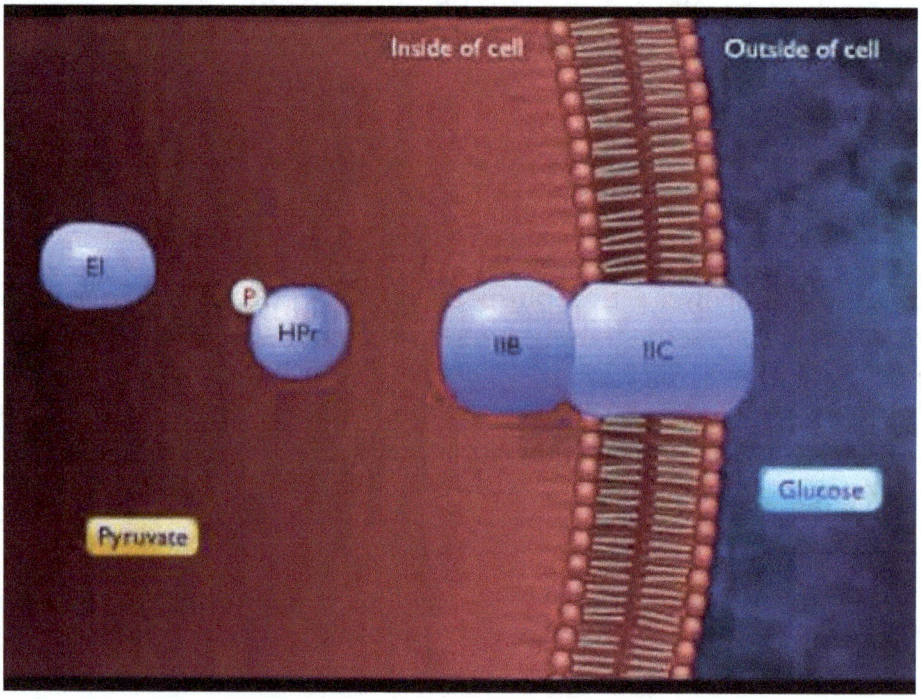

Group translocation: Bacterial PTS transport.

PTS's are widely distributed in prokaryotes. Aerobic bacteria lack PTS's. Genera *Escherichia, Salmonella, Staphylococcus* and other facultative anaerobic bacteria have phosphotransferase systems; some obligate anaerobic bacteria (*Clostridium*) also have PTS's. Many carbohydrates are transported by these systems. *E. coli* takes up glucose, fructose, mannitol, sucrose, N-acetylglucosamine, cellobiose and other carbohydrates by group translocation.

Iron Uptake:

All microorganisms require iron for use in cytochromes and many enzymes. Iron uptake is made difficult by the extreme insolubility of ferric ion (Fe^{3+}) and its derivatives, which leave little free, iron available for transport. Many bacteria and fungi have overcome this difficulty by secreting siderophores. Siderophores – are low molecular weight molecules that are able to complex with ferric ion and supply it to the cell. These are either hydroxamates or phenolates-catecholates. Ferrichrome is a hydroxamate produced by many fungi; enterobactin is the catecholate formed by *E. coli.* Microorganisms secrete sidereophores when little iron is available in the medium. Once the iron-siderophore complex has reached the cell surface, it binds to a siderophore receptor protein. The iron is either released to enter the cell directly or the whole iron-siderophore complex is transported inside by an ABC transporter. In *E. coli*, the siderophore receptor is in the outer membrane of the cell envelope; when the iron reaches the periplasmic space, it moves through the plasma membrane with the aid of the transporter. After the iron has entered the cell, it is reduced to the ferrous form (Fe^{2+}). Iron is so crucial to microorganisms that many use more than one route of iron uptake to ensure an adequate supply.

Culture Media

In the laboratory, bacteria are normally grown or cultured in either liquid medium, in flasks, bottles or large culture vessels called fermenters or solid medium in Petri dishes which are round, normally plastic or glass dishes. Introduction of microbes into or onto these media is called inoculation. The nature of the medium depends on the microorganism's natural environment because its nutrient requirements reflect its natural surroundings. The culture medium must contain all the nutrients that the microorganism requires for growth like water, a source of energy, carbon, nitrogen, essential inorganic ions and a number of trace elements. Bacteria that can synthesize all they require from the basic ingredients are called Prototrophs – and most microorganisms that survive in the outside environment can do this. Microbes that have become adapted to life in a situation rich with nutrients such as human body may require other growth factors like vitamins, amino acids or nitrogenous bases to be provided. These are called auxotrophs.

The media that are used in microbiology laboratories to culture bacteria are referred to as artificial media or synthetic media, because they do not occur naturally; rather they are prepared in the laboratory. There are number of ways of categorizing the media, one way is to classify based on whether the exact contents of the media are known or not.

Chemically defined media : Is one in which all the ingredients are known; and was prepared in the laboratory by adding a certain number of grams of each of the components (Carbohydrates, amino acids, salts etc). Particularly photolithoautotrophs such as Cyanobacteria and eukaryotic algae can be grown on relatively simple media containing CO_2 as a carbon source (often added as sodium carbonate or bicarbonate), nitrate or ammonia as a nitrogen source, sulfate, phosphate and a variety of minerals. Chemoorganoheterotrophs can be grown in a defined media with glucose as a carbon source and an ammonium salt as a nitrogen source.

Complex media : contains undefined ingredients and the exact contents are not known. Complex medium may be sufficiently rich and complete to meet the nutritional requirements of many different microorganisms. They contain undefined components like peptones, meat extract and yeast extract. Peptones – are proteins hydrolysates prepared by partial proteolytic digestion of meat, casein, soyameal, gelatin and other protein sources (Carbon, energy and nitrogen). Beef extract – aqueous extracts from lean beef and contain amino acids, peptides, nucleotides, organic acids and minerals and vitamins.Yeast extract – from brewer's yeast and contain an excellent source of B vitamins, nitrogen and carbon compounds. Three commonly used complex media are:

1. Nutrient broth – Peptone (5.0g/L), Beef Extract (3.0g/L).

2. Tryptic soya broth – Tryptone (enzyme digest of casein), Peptone (enzyme digest of soybean meal), glucose, NaCl and K_2HPO_4.

3. MacConkey Agar – Pancreatic digest of gelatin, casein, peptic digest of animal tissue, bile salts mixture, NaCl, neutral red, crystal red and agar.

These media are routinely used for cultivation of bacteria in the laboratory and particularly useful for cultivation of bacteria whose growth requirements have not been defined.

Culture media can also be categorized as liquid or solid. Liquid media (also called broths) are contained in tubes, flasks and fermenters. Solid media are prepared by adding agar to liquid medium and then pouring the media into tubes or Petridishes where the media will solidify. Agar is a complex polysaccharide that is obtained from red marine algae. Other solidifying agents are gelatin, silica. A 1% or 2% agar can be used to solidify, but 1.5% is the most commonly used.

Enriched Medium : Is a broth or solid medium containing a rich supply of special nutrients that promotes the growth of fastidious organisms (that have complex nutrition and environmental requirements). It is usually prepared by adding extra nutrients to a medium called nutrient agar. Blood agar (nutrient agar + 5% sheep red blood cells): It is bright red in color and distinguishes between the hemolytic and non-hemolytic bacteria (*Streptococci* and other pathogens).

Chocolate agar (nutrient agar + powdered hemoglobin): It is brown in color and is considered more enriched than blood agar as hemoglobin is more readily accessible in chocolate agar. Pathogens like *Neisseria gonorrhoeae* and *Haemophillus influenza*, which will not grow on blood agar, can be cultured.

Types of Media:

Selective Media: Are designed to suppress the growth of unwanted bacteria and encourage the growth of the desired microbes.

Examples: MacConkey agar inhibits growth of gram positive bacteria and thus is selective for gram-negative bacteria. Phenylethyl alcohol (PEA) agar and Colistinalidixic acid (CAN) agar are selective for gram positive bacteria.

Manitol salt agar contains a concentration of 7.5% NaCl that is quite inhibitory to most human pathogens. Bile salts, a component of fecus, inhibit most gram positive bacteria while permitting many gram negative rods to grow. This media is used for selecting intestinal pathogens which contain bile salts. Dyes such as methylene blue and crystal violet also inhibit certain gram positive bacteria (Staphylococcus can produce acid from mannitol and turn the phenol red dye to bright yellow.

A medium containing acetate as a carbon source would be selective for organisms that grow on acetate. Sabaraud's dextrose agar is used to isolate fungi and cellulose for cellulose digesting bacteria.

Differential media : Makes it easier to distinguish colonies of the desired organism from other colonies growing on the same plate.

MacConkey agar is also a differential and selective medium to distinguish between lactose fermenting organisms (e.g, *E. coli*) and other non-fermenters (e.g, *Shigella* sp.). Lactose in the medium is fermented by *E. coli* producing acid which causes an indicator dye to change color to red and colonies that do not ferment lactose are white. Dyes can be used as differential agents because many of them are pH indicators that change color in response to the production of an acid or base. MacConkey agar contains neutral red, a dye that is yellow when neutral and pink or red when acidic.

Mannitol salt agar is used to screen for *Staphylococcus aureus,* and it turns the originally pink medium to yellow due to its ability to ferment mannitol. In a sense, blood agar is also differential because it is used to determine the hemolytic and non-hemolytic bacteria.

Blood agar is both differential and an enriched one. It distinguishes between hemolytic and non-hemolytic bacteria. Hemolytic bacteria (eg., many *Streptococci* and *Staphylococci*) produce clear zones around their colonies because of red blood cell destruction.

Isolation of pure cultures:

In natural habitats, microorganisms usually grow in complex, mixed populations containing several species. One needs a pure culture, a population of cells arising from a single cell, to characterize an individual species. Robert Koch introduced pure culture techniques.

Spread plate and streak plate:

Spread plate - Mixture of cells is spread out on an agar surface so that every cell grows into a completely separate colony, each colony representing a pure culture. A small volume of around 30 to 300 cells (mixed) is transferred to the center of the plate and spread evenly over the surface with a sterile bent-glass rod.

Streak plate – the microbial mixture is transferred to the edge of an agar plate with an inoculating loop or swab and then streaked out over the surface in one of several patterns. In both these techniques, successful isolation depends on spatial separation of single cells.

The pour plate : Extensively used with bacteria and fungi. The original sample is diluted several times to reduce the microbial population sufficiently to obtain separate colonies when plating. The microbes are mixed with molten agar which has been cooled t 45° Cand then poured into pertidishes. All these techniques require petridishes (special culture dishes) after their inventor Julius Richard Petri (1887). They consist of two round halves, the top half overlapping the bottom. Each bacterium replicates to form a colony that is visible to the naked eye. A colony contains up to 109 copies of the original bacterium.

Colony morphology and growth:

Colony development on agar surfaces aids the microbiologist in identifying bacteria because individual species often form colonies of characteristic size and appearance. Structure of bacterial colonies also has been examined with the scanning electron microscope. Generally the most rapid cell growth occurs at the colony edges. Growth is much slower in the center and cell autolysis takes place in the older central portion of some colonies. These differences in growth appear due to gradients of oxygen, nutrients and toxic products within the colony. At the colony edge, oxygen and nutrients are plentiful. Bacteria growing on solid surfaces such as agar can form complex and intricate colony shapes. Nutrient diffusion and availability, bacterial chemotaxis and the presence of liquid on the surface all appear to play a role in pattern formation. These patterns vary with nutrient availability and the hardness of the agar surface.

Growth of Bacterial Cultures

Growth is defined as an increase in cellular constituents which leads to a rise in cell number. As we are aware, microorganisms reproduce by binary fission or by budding. In order to study growth, normally one follows the changes in the total population number. The cells copy their DNA almost continuously and divide again and again by the process called binary fission . Binary fission which has been described earlier a s the form of asexual reproduction in single-celled organisms by which one cell divides into two cells of the same size. Fortunately, few prokaryotic populations can sustain exponential growth for long. Environments are usually limiting in resources such as food and space. Prokaryotes also produce metabolic waste products that may eventually pollute the colony's environment. Still, you can understand why certain bacteria can make you sick so soon after infection or why food can spoil so rapidly. Refrigeration retards food spoilage not because the cold kills the bacteria on food but because most microorganisms reproduce very slowly at such low temperatures.

Fission (Biology)

In biology, fission is the division of a single entity into two or more parts and the regeneration of those parts into separate entities resembling the original. The object experiencing fission is usually a cell, but the term may also refer to how organisms, bodies, populations, or species split into discrete parts. The fission may be *binary fission*, in which a single entity produces two parts, or *multiple fission*, in which a single entity produces multiple parts.

Binary Fission

Organisms in the domains of Archaea and Bacteria reproduce with binary fission. This form of asexual reproduction and cell division is also used by some organelles within eukaryotic organisms (e.g., mitochondria). Binary fission results in the reproduction of a living prokaryotic cell (or organelle) by dividing the cell into two parts, each with the potential to grow to the size of the original.

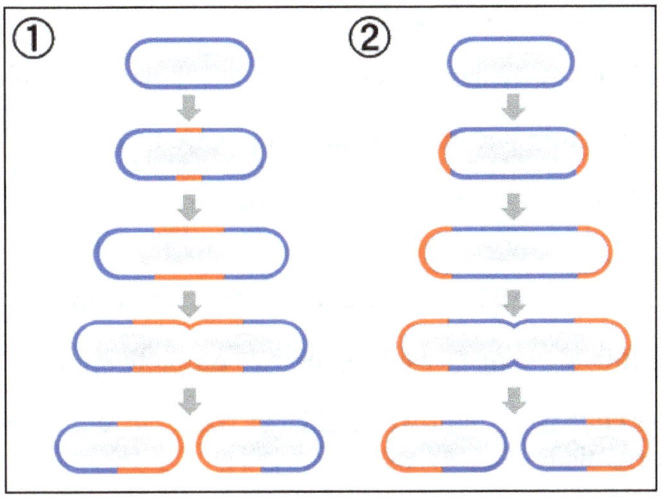

Schematic diagram of cellular growth (longation) and binary fission of bacilli. Blue and red lines indicate old and newly generated bacterial cell wall, respectively. (1) growth at the center of bacterial body. e.g. *Bacillus subtilis, E. coli*, and others. (2) apical growth. e.g. *Corynebacterium diphtheriae*.

Fission of Prokaryotes

Unlike the process of mitosis used by eukaryotic cells, binary fission takes place without the formation of a spindle apparatus in the cell. The single DNA molecule first replicates, then attaches each copy to a different part of the cell membrane. When the cell begins to pull apart, the replicated and original chromosomes are separated. The consequence of this asexual method of reproduction is that all the cells are genetically identical, meaning that they have the same genetic material (barring random mutations).

Process of Bacterial Fission

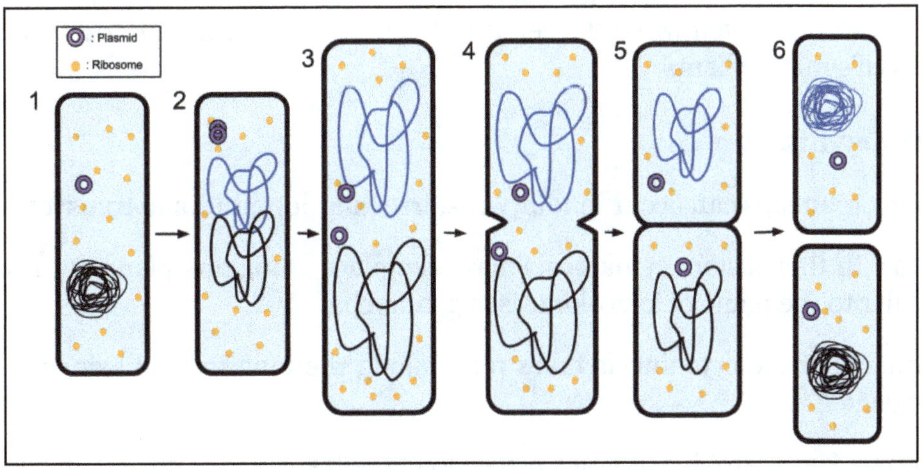

Binary fission in a prokaryote.

The process of binary fission in bacteria involves the following steps. First, the cell's DNA is replicated. The replicated DNA copies then move to opposite poles of the cell in an energy-dependent process. The cell lengthens. Then, the equatorial plate of the cell constricts and separates the plasma membrane such that each new cell has exactly the same genetic material.

More specifically, the following steps occur:

1. The bacterium before binary fission is when the DNA is tightly coiled.
2. The DNA of the bacterium has uncoiled and duplicated.
3. The DNA is pulled to the separate poles of the bacterium as it increases size to prepare for splitting.
4. The growth of a new cell wall begins to separate the bacterium.
5. The new cell wall fully develops, resulting in the complete split of the bacterium.
6. The new daughter cells have tightly coiled DNA rods, ribosomes, and plasmids; these are now brand new organisms.

Speed of Bacterial Fission

Binary fission is generally rapid though its speed varies between species. For *E. coli*, cells typically divide about every 20 minutes at 37 °C. Because the new cells will, in turn, undergo binary fission on their own, the time binary fission requires is also the time the bacterial culture requires to double in the number of cells it contains. This time period can therefore be referred to as the doubling time. Some species other than *E. coli* may have faster or slower doubling times: some strains of *Mycobacterium tuberculosis* may have doubling times of nearly 100 hours. Bacterial growth is limited by factors including nutrient availability and available space, so binary fission occurs at much lower rates in bacterial cultures once they enter the stationary phase of growth.

Fission of Organelles

Some organelles in eukaryotic cells reproduce using binary fission. Mitochondrial fission occurs frequently within the cell, even when the cell is not actively undergoing mitosis, and is necessary to regulate the cell's metabolism.

Types of Binary Fission

Binary fission in organisms can occur in four ways, irregular,longitudinal, transverse, oblique.

1. Irregular: In this fission, cytokinesis may take place along any plane but it is always perpendicular to the plane of karyokinesis. e.g. amoeba.
2. Longitudinal: Here cytokinesis takes place along the longitudinal axis. e.g. in flagellates like Euglena.
3. Transverse: Here cytokinesis takes place along the transverse axis. e.g. in *ciliate* protzoans like paramecium.
4. Oblique: In this type of binary fission cytokinesis occurs obliquely. e.g. ceratium.

Multiple Fission

Fission of Protists

Multiple fission at the cellular level occurs in many protists, e.g. sporozoans and algae. The nucleus of the parent cell divides several times by mitosis, producing several nuclei. The cytoplasm then separates, creating multiple daughter cells.

Some parasitic, single-celled organisms undergo a multiple fission-like process to produce numerous daughter cells from a single parent cell. Isolates of the human parasite *Blastocystis hominis* were observed to begin such a process within 4 to 6 days. Cells of the fish parasite *Trypanoplasma borreli* have also been observed participating in both binary and multiple fission.

Fission of Apicomplexans

In the apicomplexans, a phylum of parasitic protists, multiple fission, or schizogony, is manifested either as merogony, sporogony or gametogony. Merogony results in merozoites, which are multiple daughter cells, that originate within the same cell membrane, sporogony results in sporozoites, and gametogony results in microgametes.

Fission of Green Algae

Green algae can divide into more than two daughter cells. The exact number of daughter cells depends on the species of algae and is an effect of temperature and light.

Multiple Fission of Bacteria

Most species of bacteria primarily undergo binary reproduction. Some species and groups of bacteria may undergo multiple fission as well, sometimes beginning or ending with the production of spores. The species *Metabacterium polyspora*, a symbiont of guinea pigs, has been found to produce multiple endospores in each division. Some species of cyanobacteria have also been found to reproduce through multiple fission.

Plasmotomy

Some protozoans reproduce by yet another mechanism of fission called as plasmotomy. In this type of fission, a multinucleate adult parent undergoes cytokinesis to form two multinucleate(or coenocytic) daughter cells. The daughter cells so produced undergo karyokinesis, further.

Such type of reproduction are shown by opalina and pelomyxa.

Clonal Fragmentation

Fragmentation in multicellular or colonial organisms is a form of asexual reproduction or cloning where an organism is split into fragments. Each of these fragments develop into mature, fully grown individuals that are clones of the original organism. In echinoderms, this method of reproduction is usually known as fissiparity.

Population Fission

Any splitting of a single population of individuals into discrete parts may be considered fission. A population may undergo fission for a variety of reasons, including migration or geographic isolation. Because the fission leads to genetic variance in the newly isolated, smaller populations, population fission is a precursor to speciation.

Growth Curve

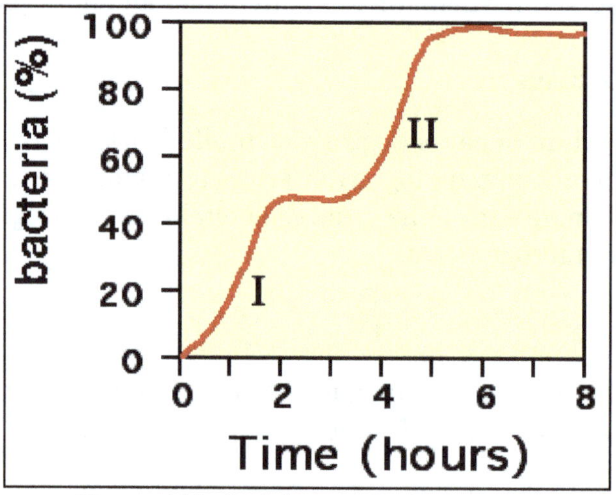

Figure: A bi-phasic bacterial growth curve.

A growth curve is an empirical model of the evolution of a quantity over time. Growth curves are widely used in biology for quantities such as population size or biomass (in population ecology and demography, for population growth analysis), individual body height or biomass (in physiology, for growth analysis of individuals). Values for the measured property can be plotted on a graph as a function of time.

Bacterial Growth

In this example the number of bacteria present in a nutrient-containing broth was measured during the course of an 8-hour cell growth experiment. The observed pattern of bacterial growth is bi-phasic because two different sugars were present, glucose and lactose. The bacteria prefer to consume glucose (Phase I) and only use the lactose (Phase II) after the glucose has been depleted. Analysis of the molecular basis for this bi-phasic growth curve led to the discovery of the basic mechanisms that control gene expression.

Cancer Cell Growth

Cancer research is an area of biology where growth curve analysis plays an important role. In many types of cancer, the rate at which tumors shrink following chemotherapy is related to the rate of tumor growth before treatment. Tumors that grow rapidly are generally more sensitive to the toxic effects that conventional anticancer drugs have on the cancer cells. Many conventional anticancer drugs (for example, 5-Fluorouracil) interfere with DNA replication and can cause the death of cells that attempt to replicate their DNA and divide. A rapidly growing tumor will have more actively dividing cells and more cell death upon exposure to such anticancer drugs.

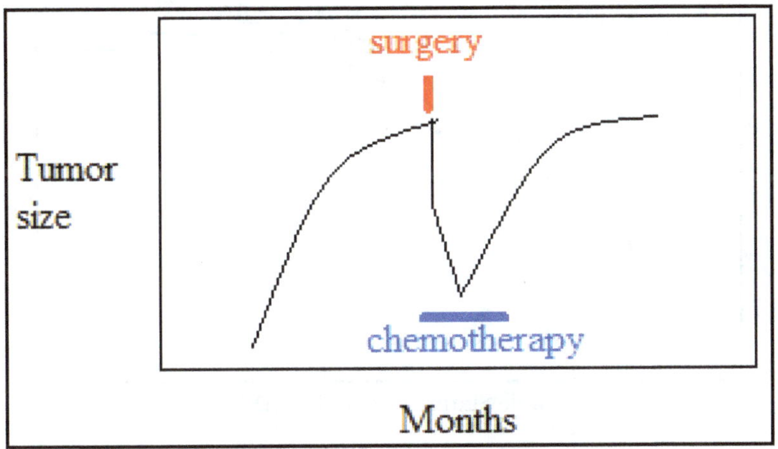

Figure. Tumor growth curve.

In the example shown in Figure, a tumor is found after the cell growth rate has slowed. Most of the cancer cells are removed by surgery. The remaining cancer cells begin to proliferate rapidly and cancer chemotherapy is started. Many tumor cells are killed by the chemotherapy, but eventually some cancer cells that are resistant to the chemotherapy drug begin to grow rapidly. The chemotherapy is no longer useful and is discontinued.

The Growth of Children

Children who fall significantly below the normal range of growth curves for body height can be tested for growth hormone deficiency and might be treatable with hormone injections.

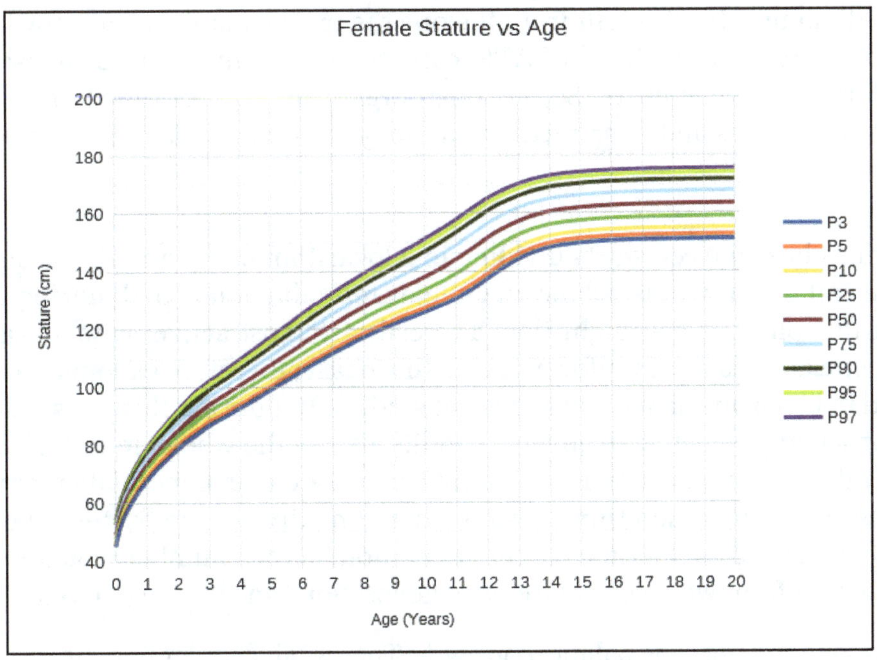

Female Stature vs Age (US CDC).

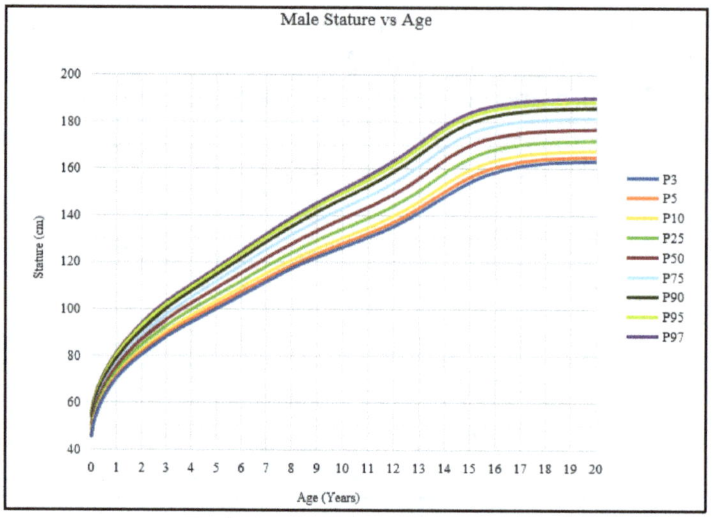

Male Stature vs Age (US CDC).

Exponential Growth

Some growth curves for certain biological systems display periods of exponential growth. Typically, periods of exponential growth are of limited duration due to depletion of some rate-limiting resource.

Lag Phase

Lag Phase: As the cells are introduced into the new medium, no immediate increase in cell number occurs. During this phase, the cells are undergoing a period of intense metabolic activity involving synthesis of enzymes and various other molecules required to divide in the coming phase. This phase can vary considerably in length depending on the nature of the medium and the microorganism. The medium may be different from the one the microorganism was growing in previously. The cells may be old and depleted of ATP, essential cofactors and ribosomes; these must be synthesized before growth can begin. So, the microorganism requires time to recover and young, vigorously growing cultures and fresh medium are to be used for the lag phase to be short.

Log Phase

Log Phase: In this phase the cell starts dividing in a logarithmic way and this is also called as exponential phase and the growth is balanced. Cellular reproduction is high during this period and the plot during this phase is a straight line. The cells are most active metabolically during this phase and the population is most uniform during this phase; therefore exponential phase cultures are usually used in biochemical and physiological studies. But during this phase, the microorganisms may be particularly sensitive to adverse conditions. On the whole, in this phase the cells are growing and dividing and increasing in cell number.The rate of exponential growth of a bacterial culture is expressed as generation time , also the doubling time of the bacterial population. Generation time (G) is defined as the time (t) per generation/ n (n = number of generations). Hence, G=t/n is the equation from which calculations of generation time can be derived.

Exponential phase or log phase is balanced growth. That is, all cellular constituents are manufactured at constant rates relative to each other. If nutrient levels or other environmental conditions change, unbalanced growth results. The individual cells may take slightly longer than others to go from lag phase to the log phase, and they do not all divided precisely together. If they divided together and the generation time is same, the number of cells in a culture would increase in a stair – step pattern, exactly doubling every 20 min or a particular time – a hypothetical situation called Synchronous growth. In an actual culture, each cell divides sometime during the 20 min generation time, with about 1/20 cells dividing each minute – a natural situation called nonsynchronous growth or asynchronous growth which appears as a smooth line, not as steps.

Organisms in a tube of culture medium can maintain logarithmic growth for only a limited time. As the number of organisms' increases, nutrients are used up, metabolic wastes accumulate, living space may become limiting factor and aerobes suffer from oxygen depletion.

Stationary Phase

Stationary Phase: Exponential growth cannot be continued forever in a batch culture (e.g. a closed system such as a test tube or flask). Population growth is limited by one of three factors: 1. exhaustion of available nutrients; 2. accumulation of inhibitory metabolites or end products; 3. exhaustion of space, in this case called a lack of "biological space".During the stationary phase, if viable cells are being counted, it cannot be determined whether some cells are dying and an equal number of cells are dividing, or the population of cells has simply stopped growing and dividing. The stationary phase, like the lag phase, is not necessarily a period of quiescence. Bacteria that produce secondary metabolites, such as antibiotics, do so during the stationary phase of the growth cycle (Secondary metabolites are defined as metabolites produced after the active stage of growth). It is during the stationary phase that spore-forming bacteria have to induce or unmask the activity of dozens of genes that may be involved in sporulation process. Starving bacteria frequently produce a variety of starvation proteins, which make the cell much more resistant to damage. They increase peptidoglycan cross-linking and cell wall strength. The Dps (DNA- binding proteins from starved cells) protein protects the DNA. Bacterial pathogens like *Salmonella typhymurium* become more virulent when starved.

Death Phase

Death Phase: Due to the conditions during the stationary phase, the death phase is seen as there is a decline in the number of viable cells. This phase also is like the log phase where the cell number is declining in a logarithmic way. The cell is said to be dead if it does not revive itself and reproduce when incubated again in a fresh medium. In this phase, the number of live cells decreases at a logarithmic rate, as indicated by the straight downward sloping diagonal line. The duration of this phase is as highly variable as the duration of log phase. Both depend primarily on the genetic characteristics of the organism.

Mathematics of Growth:

Microbial growth during the exponential phase is very important and of interest to microbiologists and the analysis applies to microorganisms dividing by binary fission. The time required by a cell to divide is called the generation time or doubling time. In the laboratory, under favorable conditions, a growing bacterial population doubles at regular intervals. Growth is by geometric progression: 1, 2, 4, 8, etc. or 2^0, 2^1, 2^2, 2^3.........2^n (where n = the number of generations). This is called exponential growth. In reality, exponential growth is only part of the bacterial life cycle, and not representative of the normal pattern of growth of bacteria in Nature. This might vary from organism to organism depending upon the environmental conditions etc. For example in *E. coli* the generation time is 20 min and hence after 20 generations a single initial cell would increase to over 1 million cells. This would require a little less than 7 hours. The population is doubling every generation; hence the increase in population is always $2^n$where n is the number of generations. The resulting population increase is exponential or logarithmic.

When growing exponentially by binary fission, the increase in a bacterial population is by geometric progression. If we start with one cell, when it divides, there are 2 cells in the first generation, 4 cells in the second generation, 8 cells in the third generation, and so on. The generation time is the time interval required for the cells (or population) to divide.

G (generation time) = (time, in minutes or hours)/n(number of generations).

$G = t/n$.

t = time interval in hours or minutes.

B = number of bacteria at the beginning of a time interval.

b = number of bacteria at the end of the time interval.

n = number of generations (number of times the cell population doubles during the time interval).

$b = B \times 2^n$ (This equation is an expression of growth by binary fission).

Solve for n:

$\log b = \log B + n\log 2$.

$n=\log b-\log B/\log 2$.

$n= \log b-\log B/ .301$.

$n = 3.3 \log b/B$.

$G = t/n$.

Solve for G.

$G= t/3.3\log b/B$.

Example: What is the generation time of a bacterial population that increases from 10,000 cells to 10,000,000 cells in four hours of growth?

$G= t/3.3\log b/B$.

$G= 240\text{minutes}/3.3\log 10^7 /10^4$.

$G= 240\text{minutes}/3.3\text{x}3$.

G = 24 minutes.

Measurement of Microbial Growth:

A number of techniques are available in order to measure growth of microbial populations. Either population number of mass may be calculated ad growth leads to increase in both.

Direct measurement of cell numbers:

Bacteria or microorganisms can be counted directly on the plate and also called as plate counting. Advantage of this method is that it measures the number of viable cells. Disadvantage is that, it is time consuming and expensive as one needs media and other conditions need to be maintained. Bacteria counted on plate counts are referred to as colony forming units as a single cell or a clump of bacterial cells can lead to a colony which contains many cells. The colonies when they are counted in plate count method are to be present sparsely for accurate counting as overcrowding can lead to incorrect counting. To solve this, one has to adapt the serial dilution method in order to get an accurate count.

Serial dilution and pour and spread plate: Supposing one has to accurately count the number of cells given in a solution, then serial dilution needs to be performed. A 1ml of the sample is taken and transferred to a tube containing 9ml of sterile water and this process can be repeated until we reach a considerable dilution (say 10^6 to 10^7). Once the original inoculum is diluted one needs to perform a pour plate or a spread plate technique in order to count the number of bacteria present in the diluted sample and then the original sample. In pour plate method the diluted sample is poured into the petriplate and then the medium which is at nearly 50°C is poured over the inoculum and mixed by gentle agitation. With this method, colonies grow within the nutrient agar as well as on the surface of the agar plate. As certain disadvantages are encountered in this method like heat sensitive microorganisms might not grow and also bacteria when they grow within the nutrient medium might not be useful for diagnostic purposes. In order to avoid these problems, spread plate method is mostly used. A 0.1ml of the diluted sample is added to the surface of the nutrient medium and spread uniformly with the help of a glass spreader and after incubation, the colonies can be counted and the concentration of the bacterial cells in the original sample is calculated as follows:

Number of bacteria/ml = Number of colonies on plate x reciprocal of dilution of sample.

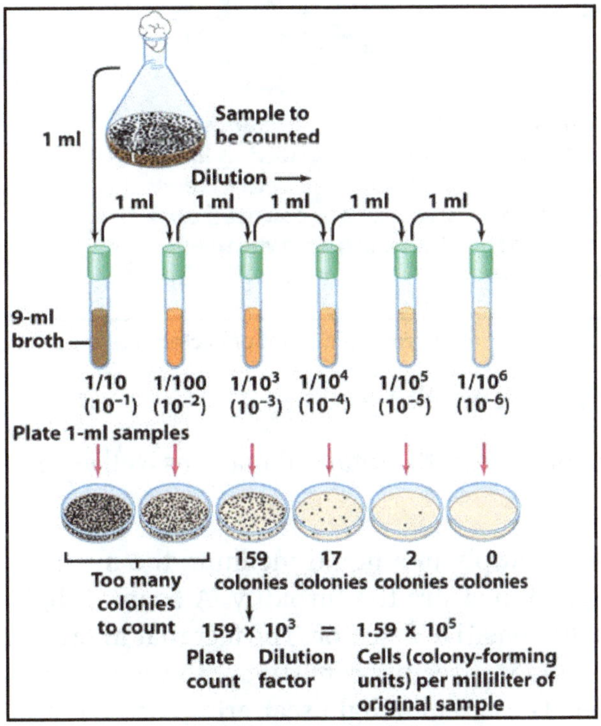

Serial dilution methodology.

Membrane Filtration: This method can be used in order to study if the quantity of the bacteria is very small as in aquatic samples like lakes, streams etc. Membranes with different pore sizes are used to trap different microorganisms. The sample is drawn through these special membrane filters and placed on an agar medium or on a pad soaked with liquid media. After incubation, the number of colonies can be counted and the number determined in the original sample. Selective media or differential media can be used for specific microorganisms. This is mostly used for analyzing aquatic samples.

Microscopic count: The Petroff-Hausser counting chamber or slide is easy, inexpensive and relatively quick method and also gives information about the size and morphology of the microorganisms. These specially designed slides have chamber of known depth with an etched grid on the chamber bottom.

Bacteria/mm³ = (bacteria/square) (25 squares)/ (50)

Bacteria can be counted by taking into account the chamber's volume and any sample dilution. The disadvantage encountered in this method is that fairly large volume is required and also it is difficult to distinguish between living and dead cells. Microorganisms of larger sizes can be counted by using electronic counters such as coulter counter; where in the number of cells in a measured volume of liquid is counted. This method gives accurate results with larger cells and is extensively used in hospital laboratories to count red and white blood cells.

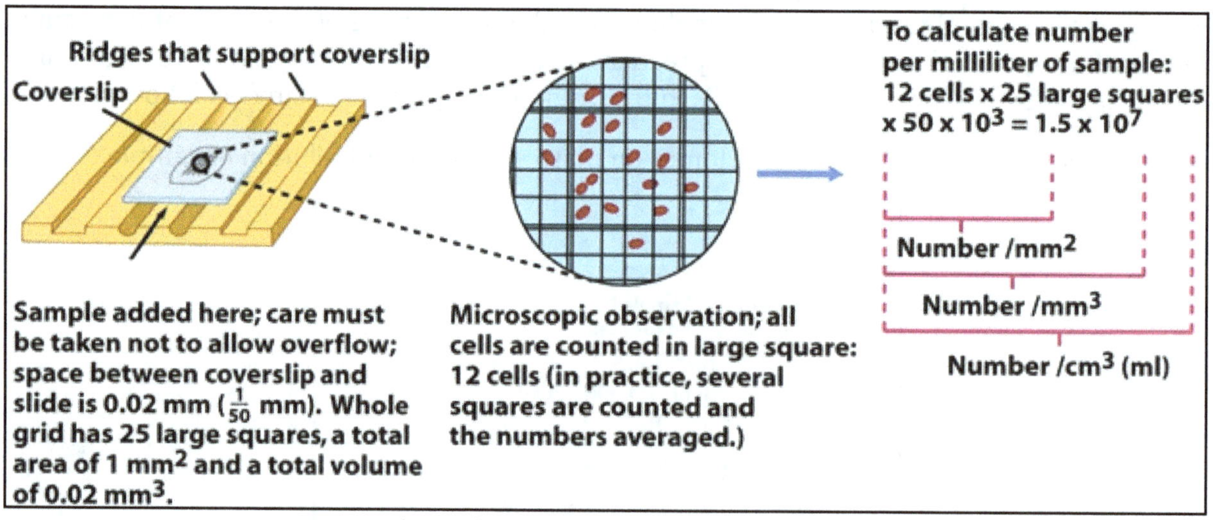

Direct microscopic count of bacterial cells.

Indirect methods of measurement of cell mass:

Population growth leads to increase in the total cell mass, as well as in cell numbers. The following methods can be used.

Turbidity: As bacteria grow/multiply in a liquid medium, the medium becomes turbid. Spectrophotometer is used in order to measure the turbidity. A beam of light is transmitted through a bacterial suspension to a light-sensitive detector. The fact that microbial cells scatter light striking them, the amount of scattering is directly proportional to the biomass of cells present and indirectly related to cell number. The extent of light scattering can be measured and is almost linearly related to bacterial concentration at low absorbance levels.

Broth culture showing turbidity.

Dry weight: This method is mostly used for filamentous bacteria and moulds. The microorganism is grown in liquid medium, filtered or centrifuged to remove extraneous material, and dried in an oven and then weighted. It is time consuming and hence not very sensitive.

Continuous culture of Microorganisms:

Batch cultures : Nutrient supplies are not renewed nor wastes removed.

Continuous cultures : continuous provision of nutrients and removal of wastes takes place. The population can be maintained in the exponential phase and at a constant biomass concentration for extended periods.

These can again be categorized into two types:

Chemostat : Where sterile medium is fed into the culture vessel at the same rate as the media containing microorganisms is removed.

Turbidostat : It has a photocell that measures the absorbance or turbidity of the cell culture in the growth vessel. The flow rate of media through the vessel is automatically regulated to maintain a predetermined turbidity or cell density.

Used in food and industrial microbiology.

Environmental Factors on Microbial Growth

The growth of microorganisms is greatly affected by the chemical and physical nature of their surroundings. An understanding of these influences aids in the control of microbial growth and the study of the ecological distribution of microorganisms. Prokaryotes are present or grow anywhere life can exist. The environments in which some prokaryotes grow would kill most other organisms. For example *Bacillus infernus* is able to live over 1.5 miles below the earth's surface without O_2 and 60°C temperature. These microorganisms which can thrive and grow in such harsh conditions are often called extremophiles.

The major physical factors which affect microbial growth are solutes and water activity, pH, temperature, oxygen level, pressure and radiation.

Solutes and Water activity: Changes in osmotic concentration of the surroundings can affect microbial growth as a selectively permeable plasma membrane separates the microorganisms from their surroundings. Microorganisms need to keep the osmotic concentration of their cytoplasm somewhat above that of the habitat by the use of compatible solutes, so that the plasma membrane is always pressed firmly against their cell wall. In a hypertonic environment, the prokaryotes increase their internal osmotic concentration through the synthesis or uptake of choline, proline, glutamic acid and other amino acids. A few prokaryotes like *Halobacterium salinarium* raise their osmotic concentration with potassium ions. The enzymes of these bacteria are altered for the requirement of high salt concentrations for normal activity. Halophiles grow optimally in the presence of NaCl or other salts at a concentration above about 0.2M. These have extensively modified the structure of their proteins and membranes rather than simply increasing the intracellular concentrations of solutes. They require higher potassium levels for stability and activity. The plasma membrane of halophiles is also stabilized by high concentration of sodium ions.

Water activity (a_w) is the amount of water available to microorganisms and this can be reduced by interaction with solute molecules (osmotic effect). Water activity is inversely related to osmotic pressure; if a solution has high osmotic pressure, it's a w is low. Microorganisms differ greatly in their ability to adapt to habitats with low water activity. In a low a w habitat, the microorganisms must expend extra effort to grow as it should maintain a high solute concentration to retain water. Such microorganisms are osmotolerant or can grow over wide range of water activity or osmotic concentration. Most of the microorganisms grow at a w =0.98 or higher.

pH: It refers to the acidity or alkalinity of a solution. It is a measure of the hydrogen ion activity of a solution and is defined as the negative logarithm of the hydrogen ion concentration.

$pH = -\log [H^+] = \log (1/H^+)$.

The pH scale ranges from 1.0 to 14.0 and most microorganisms grow vary widely from pH 0 to 2.0 at the acid end to alkaline lakes and soil that may have pH values between 9.0 and 10. The pH can affect the growth of microorganisms and each species has a definite pH growth range and pH growth optimum. Acidophiles have their growth optimum between pH 0 and 5.5; neutrophiles between 5.5 and 8.0 and alkalophiles prefer pH range of 8.5 to 11.5. Most bacteria and protozoans are neutrophiles, fungi prefer acid surroundings about pH 4 to 6; algae also seem to favour slight acidity. *Cyanidium caldarium* (algae) and archaeon *Sulfolobus acidocaldarium* are inhabitants of acidic hot springs; both grow well around pH 1 to 3 and at high temperature. Drastic changes/variations in cytoplasmic pH can harm microorganisms by disrupting the plasma membrane or inhibiting the activity of enzymes and membrane transport proteins. Prokaryotes die if the internal pH drops much below 5.0 to 5.5. External pH alterations also might alter the ionization of nutrient molecules and thus reduce their availability to the organism. The microorganism needs to maintain a neutral cytoplasmic pH and for this the plasma membrane may be relatively impermeable to protons. Neutrophiles appear to exchange potassium for protons using an antiport transport system. Extreme alkalophiles maintain their internal pH closer to neutrality by exchanging internal sodium ions for external protons. The antiport systems probably correct small variations in pH. In case of too much acidity (below 5.5 to 6.0) *S. typhimurium* and *E. coli* synthesize an array of new proteins as part of what has been called as their acidic tolerance response. If the external pH decreases to 4.5 or lower, chaperones such as acid shock proteins and heat shock proteins are synthesized. Microorganisms can change the pH of their own habitat by producing acidic or basic metabolic waste products. In order to maintain the pH, buffers are often included in the media to prevent growth inhibition. Phosphate is commonly used buffer and a good example of buffering agent. Peptides and amino acids in complex media also have a strong buffering effect.

Temperature: Temperature profoundly affects microorganisms as the most important factor influencing the effect is temperature sensitivity of enzyme-catalyzed reactions. Beyond a certain point of higher temperature, slow growth takes place and damages the microorganisms by denaturing enzymes, transport carriers and other proteins. The plasma membrane also is disrupted as lipid bilayer simply melts and the damage is such an extent that it cannot be repaired. At very low temperature, membranes solidify and enzymes don't work rapidly. In summary, when organisms are above their optimum temperature, both the function and cell structure is affected at low temperature, function is affected. The cardinal temperatures vary greatly between microorganisms. Optimum usually range from 0°C to as high as 75°C, where as microbial growth occurs at temperature extending from -20°C to over 120°C. Archaen *Geogemma barossii* grows anaerobically at 121°C. The major microbial groups differ from one another regarding their maximum growth temperature. Upper limit for protozoans is around 50°C, some algae and fungi can grow at temperatures as high as 55°C to 60°C.

Did you know: Prokaryotes have been found growing at or close of 100°C. Now thermophilic prokaryotes have been reported growing in surface of chimneys or black smokers located along rifts and ridges on the ocean floor that spew sulphide-rich super-heated vent water with temperatures above 350°C. These microbes can grow and reproduce at or above 112°C. The proteins, membranes and nucleic acids of these prokaryotes are remarkably temperature stable and provide ideal subjects for studying the ways in which macromolecules and membranes are stabilized. Some thermostable enzymes from these organisms have important industrial and scientific uses. For ex., the Taq polymerase from the thermophilic *Thermus aquaticus* is used extensively in the polymerase chain reaction.

Microorganisms are classified into five classes based on their temperature ranges for growth.

1. Psychrophiles: Microorganisms grow well at 0°C and the optimum growth temperature of 15°C or lower and maximum at around 20°C. These microorganisms are isolated from Arctic and Antarctic habitats. They have adapted to their environment in several ways. Their enzymes, transport systems and protein synthetic mechanisms function well at low temperatures. The cell membranes have high levels of unsaturated fatty acids and remain semifluid when cold. At higher than 20°C,

the psychrophiles begin to leak cellular constituents because of cell membrane disruption. Microorganisms such as *Pseudomonas, Vibrio, Alcaligenes, Bacillus, Arthrobacter, Moritella, Photobacterium* belong to this group. The psychrophilic *Chlamydomonasnivalis* turns a snowfield or glacier pink with its bright red spores.

2. Psychrotrophs or Facultative Psychrophiles: In this group many s pecies can grow ato o C to 7°C, optimum between 20°C and 30°C. The spoilage of refrigerated foods is mainly caused by microorganisms belonging to this group.

3. Mesophiles: Growth optimum around 20°C to 40°C, minimum at 15°C to 20°C and maximum at 45°C or lower. Most of the organisms fall under or within this category including human pathogens.

4. Thermophiles: The microorganisms in this group can grow at temperature of 55°C or higher, minimum is usually around 45°C and growth optima at around 55°C to 65°C. Mostly prokaryotes and a few algae and fungi belong to this group. The habitats in which they grow include, composts, self-heating haystacks, hot water lines and hot springs. Microorganisms have more heat-stable enzymes and proteins synthesis systems, which function at high temperature. Heat stable proteins have high organized, hydrophobic interiors, more hydrogen bonds and other non-covalent bonds strengthen the structure. Amino acids like proline make the polypeptide chain less flexible and chaperones also aid in folding of proteins to stabilize them. DNA also is stabilized by specific histone like proteins. The membrane lipids are also stable and tend to be more saturated, more branched and of higher molecular weight. Archaeal thermophiles have membrane lipids with ether linkages, which protect the lipids from hydrolysis at high temperatures.

5. Hyperthermophiles: Few microorganisms can grow at 96°C or above and have maximum at 100°C; and growth optima between 80°C and about 113°C. *Pyrococcus* and *Pyrpdictiumocculturn* are examples of marine hyperthermophiles found in hot floors of the sea floor.

Oxygen Concentration

Oxygen Concentration: An aerobe is an organism able to grow in the presence of atmospheric $O_2$and the ones that grow in its absence is an anaerobe. Organisms which completely are dependent on atmospheric O_2 for growth are obligate aerobes, and it serves as the terminal electron acceptor for the electron transport chain in aerobic respiration and employs it in the synthesis of sterols and unsaturated fatty acids. Organisms which do not require O_2 for growth but do grow better in its presence are called facultative anaerobes . Aerotelerant anaerobes such as Enterococcus faecalis simply ignore O2 and grow equally well whether it is present or not. Obligate anaerobes like Bacteroides, Fusobacterium, Clostridium pasteurianum, *Methanococcus, Neocallimastix*, do not tolerate O_2 at all and die in its presence. Aerotelerant and obligate anaerobes cannot generate energy through respiration and must employ fermentation or anaerobic respiration pathways for the purpose. Microaerophiles are those organisms that are damaged by the normal atmospheric levels of O_2 (20%) and require O_2 levels between the range of 2% to 16% for growth. The nature of bacterial O_2 responses can be readily determined by growing the bacteria in culture tubes filled with a solid culture medium or a special medium like thioglycollate broth, which contains a reducing agent to lower O_2 levels. Aerobic microorganisms are cultured, either the culture vessel is shaken to aerate the medium or sterile air is pumped. Anaerobic microorganisms require special anaerobic media containing reducing agents such as thioglycollate or cysteine may be used. Removing air with a vacuum pump and flushing out residual oxygen with nitrogen gas is also preferred. Co_2 and nitrogen is added to the chamber since many anaerobes require a small amount of Co_2 for best growth. The technique in which gas pak jar is used can be used.

Prokaryotes and protozoa are found arranged among all the 5 types of microorganisms. Fungi are normally aerobic, but species particularly among yeasts, are facultative anaerobes. Algae are almost always obligate aerobes. The different relationships with O_2 appear due to several factors, including the inactivation of proteins and the effect of toxic O_2 derivatives. Enzymes can be inactivated when sensitive groups like sulfydryls are oxidised. A notable example is the nitrogen fixation enzymes nitrogenise which is very O_2 sensitive.

Oxygen accepts electrons and is readily reduced because its two outer orbital electrons are unpaired. The reduction products such as superoxide radical, hydrogen peroxide and hydroxyl radical can be resulted by flavoproteins, several other cell constituents and radiation.

$$O_2 + e^- \rightarrow O_{2^-}^{\bullet}$$
$$O_{2^-}^{\bullet} + e^- + 2H^+ \rightarrow H_2O_2$$
$$H_2O_2 + e^- + H^+ \rightarrow H_2O + OH^{\bullet}$$

These are extremely toxic because they are powerful oxidizing agents and rapidly destroy cellular constituents. As microorganisms can be killed, they need to protect themselves from such oxygen products. Microorganisms possess enzymes that afford protection against toxic O_2 products. Obligate aerobes and facultative anaerobes usually contain the enzymes superoxide dismutase (SOD) and catalase, which catalyze the destruction of superoxide radical and hydrogen peroxide respectively. Peroxidase also can be used to destroy hydrogen peroxide.

$$2O_{2^-}^{\bullet} + 2H^+ \rightarrow O_2 + H_2O_2 \text{ (Superoxide dismutase)}$$
$$2H_2O_2 \rightarrow 2H_2O \rightarrow O_2 \text{ (Catalase)}$$
$$H_2O_2 + NADH + H+ \rightarrow 2H_2O + NAD^+ \text{ (Peroxidase)}$$

Aerotolerant microorganisms may lack catalase but almost always have superoxide dismutase. Strict anaerobes lack either of the enzymes or have them in very low concentrations and therefore cannot tolerate O_2. Aerobic microorganisms can be grown in an aerated medium which is aerated by shaking the vessel or sterile air is pumped. But for anaerobes, O_2 must be excluded from the medium. Reducing agents such as thioglycollate or cysteine can be used to flush out O_2. Nitrogen gas can also be used to eliminate O_2 with a vacuum pump, and sometimes CO_2 and nitrogen is added to the chamber, as anaerobes require a small amount of CO_2 for best growth and also gas pack jar can also be used.

Pressure: Most microorganisms always are subjected to pressure of 10 atmospheres (atm). The hydrostatic pressure can reach to 600 to 1100 atm in the deep sea with temperature about 2°C to 3°C. Organisms can survive and adapt at these extreme conditions and many are barotolerant, increased pressure does adversely affect them but not as much as it does to nontolerant bacteria. The barophillic organisms are those growing in the guts of deep sea invertebrates such as amphipods and holothurians and grow more rapidly at high pressures. These bacteria may play an important role in nutrient recycling in the deep sea. Bacterial genera of *Photobacteria, Shewanella, Colwellia* are barophiles. Some members of the Archaea are thermophiles for example *Pyrococcus spp., Methanococcus janaschii.*

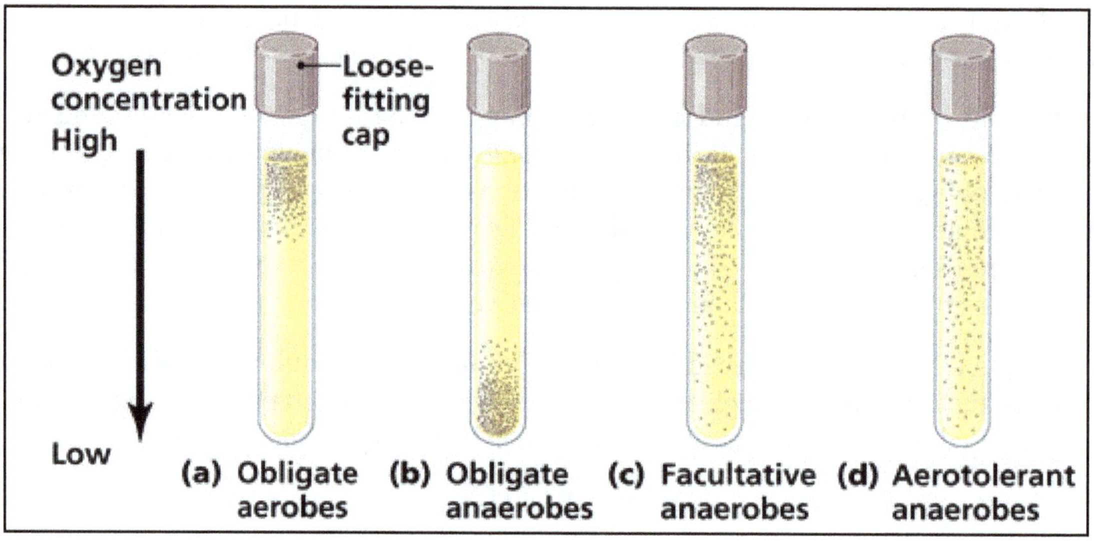

Oxygen requirements in bacteria.

Pressure:

Most organisms on land or on the surface of water is always subjected to a pressure of 1 atm. The hydrostatic pressure can reach 600 to 1100 atm in the deep sea. Despite these extremes, bacteria survive and adapt. Many are barotolerant. Some bacteria in the gut of deep sea invertebrates such as amphipods and holothurians are truly barophilic and grow more rapidly at high pressures (Ex. *Photobacterium, shewanella, Colwellia*).

Radiation:

Electromagnetic radiation of various types bombards our world. As the wavelength of electromagnetic radiation decreases, the energy of the radiation increases – gamma rays and X rays are much more energetic than visible light or infrared waves. Sunlight is the major source of radiation on the earth. It includes visible light, ultraviolet radiation, infrared rays and radio waves. Most life is dependent on the ability of photosynthetic organisms to trap the light energy of the sun as visible light. Many forms of electromagnetic radiation are very harmful to microorganisms. Ionizing radiation, radiation of very short wavelength or high energy can cause atoms to lose electrons or ionize. The two major forms of ionizing radiation, X rays which are artificially produced and gamma rays which are emitted during radioisotope decay. Low levels of ionizing radiation will produce mutations, higher levels are directly lethal. Some prokaryotes like *Deinococcus radiodurans* and bacterial endospores are resistant and can cause a variety of changes in cells like; it breaks hydrogen bonds, oxidises double bonds, destroys ring structures and polymerizes some molecules. Oxygen enhances these destructive e effects, probably through the generation of hydroxyl radicals (OH.). Destruction of DNA is the most important cause of death of microorganisms. Ultraviolet radiation kills all kinds of microorganisms due to its short wavelength (approximately 10 to 400 nm) and high energy. The most lethal UV radiation has a wavelength of 260 nm, the wavelength most effectively absorbed by DNA. Formation of thymine dimmers in DNA is the primary mechanism of UV damage; these dimmers inhibit DNA replication and function. This damage is repaired by photo reactivation, where blue light is used by a photo reactivating enzyme (photolyase) to split the thymine dimmers. Dark reactivation, where a short sequence containing the thymine dimmers can also be excised and replaced in the absence of light. Damage can also be repaired by the recA protein in recombination repair and SOS repair.

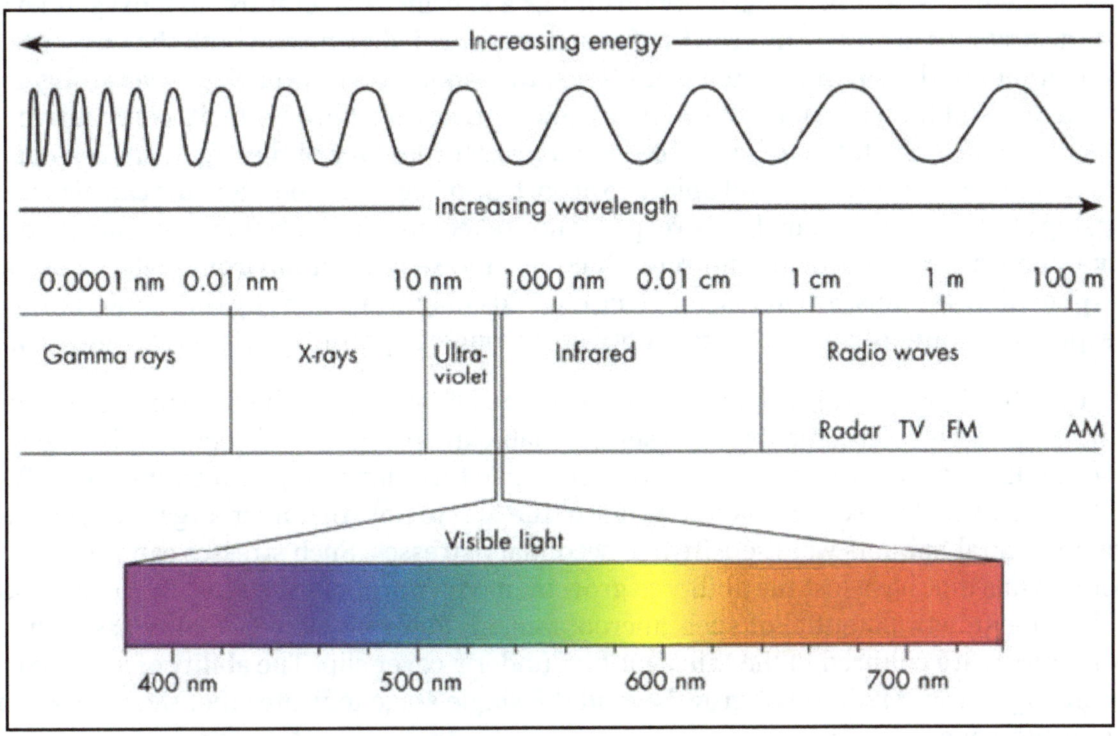

Electromagnetic spectrum.

Visible light – immensely beneficial becuas eit is the source of energy for photosynthesis. Visible light when present in sufficieint intensity can damage or kill microbial cells. Pigments called photosensitizers and O_2 are required. All microorganisms possess pigments like chlorophyll, bacteriochlorophull, cytochromes and flavins which absorb light energy, become exited or activated and act as photosensitizers. The excited photosensitizer (P) transfers its energy to O_2 generating singlet oxygen (‘O_2).

$$\text{P (light)} \rightarrow \text{P(activated)}$$

$$\text{P(activated)} + O_2 \rightarrow \text{P} + {}^{\text{‘}}O_2$$

Singlet oxygen is very reactive, powerful oxidizing agent that will quickly destroy a cell. Many microorganisms that are airborne or live on exposed surface use carotenoid pigments for protection against photooxidation. Carotenoids effectively quench singlet oxygen that is absorbing energy from singlet oxygen and convert it back into the unexcited ground state.

Microbial Growth in Natural Environments:

Microorganisms in a particular location are exposed to many overlapping gradients of nutrients and various other environmental factors. Microorganisms will grow in ‘microenvironments’ until an environmental or nutritional factor limits growth. Liebig’s law of minimum states that the total biomass of an organism will be determined by the nutrient present in the lowest concentration relative to the organisms requirements. Shelford's law of tolerance states that there are limits to environmental factors, below and above which a microorganism cannot survive and grow, regardless of the nutrient supply. Each organism has a specific temperature, pH, oxygen level and hydrostatic pressure to grow. The growth of a microorganism depends on both the nutrient supply and its tolerance of the environmental conditions. In response to low nutrient levels (oligotrophic environments) and intense competition, many microorganisms become more competitive in nutrient capture and exploitation of available resources. Morphology of the organism can change in order to increase its surface area and ability to absorb nutrients. Microorganisms can also undergo step by step shut down of metabolism except for housekeeping genes. Natural substances can also directly inhibit microbial growth and reproduction in low-nutrient environments. These agents include phenolics, tannins, ammonia, ethylene etc. This may be a means by which microorganisms avoid expending limited energy resources until an adequate supply of nutrients becomes available.

Growth of natural prokaryotic populations outside the laboratory can be determined by counting the number of viable microorganisms present. A viable microorganism is one that is able to grow actively, resulting in the formation of a colony or visible turbidity in liquid medium. John R. Postgate of University of Sussex in England was one of the first to note that microorganisms stressed by survival in natural habitats were sensitive to secondary stresses. Such stresses can produce viable microorganisms that have lost the ability to grow on media normally used for their cultivation. To determine the growth potential of such microorganism, Postgate micro viability assay, in which microorganisms are cultured in the thin agar film under a cover slip. The ability of a cell to change its morphology, even if it does not grow beyond the single stage indicates that the microorganism does show life signs.

The situation in natural environments with mixed populations is much more complex. Often only 1 to 10% of observable cells are able to form colonies. The microbiologist is attempting to grow microorganisms that perhaps never have been cultured or characterized. At present, molecular techniques involving PCR amplification and small subunit ribosomal RNA analysis are increasingly used to analyze diversity of uncultured microbial population.

Quorum sensing:

Bacteria can communicate with one another and behave cooperatively. Quorum sensing or autoinduction is a phenomenon in which bacteria monitor their own population density through sensing the levels of signal molecules, sometimes called autoinducers because they can stimulate the cell that releases them. The concentration of these signal molecules increases along with the bacterial population until it rises to a specific threshold and signals the bacteria that the population density has reached a critical level or quorum. The bacteria then begin expressing sets of quorum-dependent genes. Quorum sensing has been found among both gram-negative and gram-positive bacteria. Quorum sensing was first discovered in gram-negative bacteria and is best understood in these microorganisms. The most common signals in gram-negative bacteria are acyl-homoserine lactones (HSLs). These diffuse into the target cell and once they reach a sufficiently high level, acyl HSLs bind to specific receptor proteins and trigger conformational change, they bind to target sites on the DNA and stimulate transcription of quorum sensitive genes. The genes needed to synthesize acyl HSLs is also produced frequently, thus amplifying the effect by the production and release of more auto inducer molecules.

Examples in gram negative bacteria are 1) bioluminescence production by *Vibrio fisheri* 2) *Pseudomonas aerogionosa* synthesis and release of virulence factors 3) Conjugal transfer of genetic material by *Agrobacterium tumefaciens* and 4) antibiotic production by *Erwiniacarotovora* and *Pseudomonas aureofaciens* . Gram positive bacteria also regulate activities by quorum sensing often using an oligopeptide pheromone signals. Good examples are mating in *Enterococcus faecalis*, competence induction in *Streptococcus pneumonia*, and stimulation of sporulation by *Bacillus subtilis* , and production of many toxins and other virulence factors by *Staphylococcus aureus*. Another important function of quorum sensing is to promote the formation of mature biofilms by the pathogen, *Pseudomonas aeroginosa* , as it may play a role in cystic fibrosis. Quorum sensing is an example of what might be called multicellular behaviour in that many individual cells communicate and coordinate their activities to act as a unit. Other examples of such complex behaviour are pattern formation in colonies and fruit body formation in the Myxobacteria.

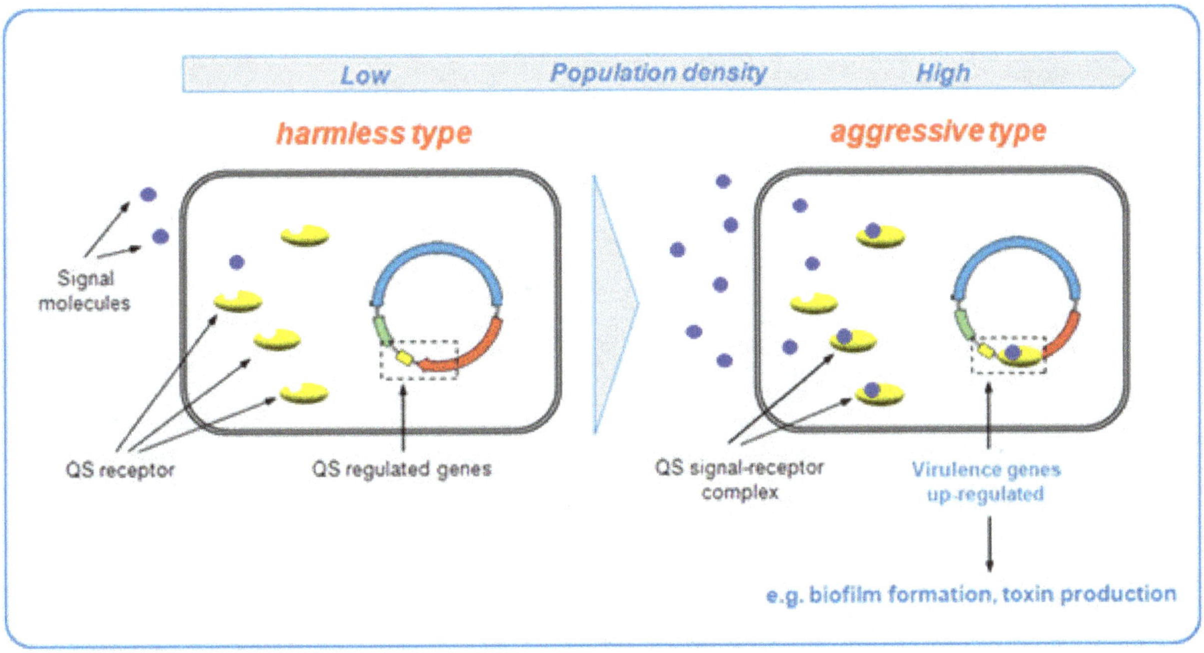

Quorum sensing.

References

- Britannica Educational Publishing (2011). Fungi, Algae, and Protists. The Rosen Publishing Group. ISBN 978-1-61530-463-9. Retrieved 06, March 2020
- Sezonov, G.; Joseleau-Petit, D.; D'Ari, R. (28 September 2007). "Escherichia coli Physiology in Luria-Bertani Broth". Journal of Bacteriology. 189 (23): 8746–8749. doi:10.1128/JB.01368-07
- P.Puranik, Asha Bhate (2007). Animal Forms And Functions: Invertebrata. Sarup & Sons. ISBN 978-81-7625-791-6. Retrieved 12, January 2020
- Rokney, Assaf; Shagan, Merav; Kessel, Martin; Smith, Yoav; Rosenshine, Ilan; Oppenheim, Amos B. (September 2009). "E. coli Transports Aggregated Proteins to the Poles by a Specific and Energy-Dependent Process". Journal of Molecular Biology. 392 (3): 589–601. doi:10.1016/j.jmb.2009.07.009. PMID 19596340

- Hubbell, S. (2003). "Modes of speciation and the lifespans of species under neutrality: a response to the comment of Robert E. Ricklefs.". Oikos. 100 (1): 193–199. doi:10.1034/j.1600-0706.2003.12450.x
- Bišová, K.; Zachleder, V. (17 January 2014). "Cell-cycle regulation in green algae dividing by multiple fission" (PDF). Journal of Experimental Botany. 65 (10): 2585–2602. doi:10.1093/jxb/ert466. PMID 24441762. Retrieved 09, May 2020
- Angert, Esther R. (March 2005). "Alternatives to binary fission in bacteria" (PDF). Nature Reviews Microbiology. 3 (3): 214–224. doi:10.1038/nrmicro1096. PMID 15738949. Retrieved 21, July 2020

Chapter 5

Concepts and Theories

There are several theories propounded in the field of microbiology. The Germ Theory of Disease is a widely accepted theory which states that microorganisms can be the cause of certain diseases. The Fermentation Theory and Hologenome Theory of Evolution are two other significant theories. This chapter deals with each theory in a manner that is easily understandable.

Germ Theory of Disease

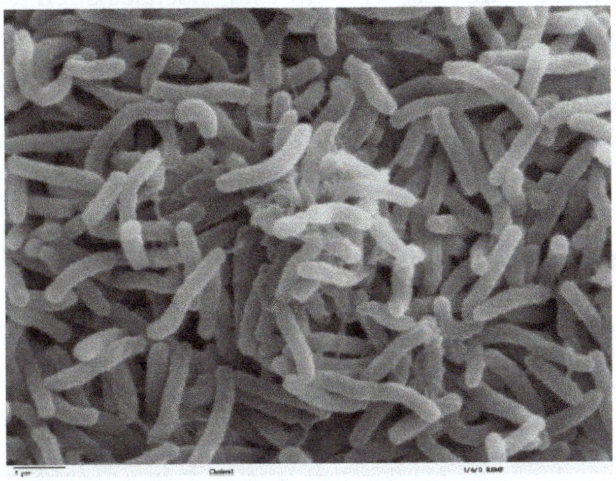

Scanning electron microscope image of *Vibrio cholerae*. This is the bacterium that causes cholera.

The germ theory of disease states that some diseases are caused by microorganisms. These small organisms, too small to see without magnification, invade humans, animals, and other living hosts. Their growth and reproduction within their hosts can cause a disease. "Germ" may refer to not just a bacterium but to any type of microorganisms, especially one which causes disease, such as protist, fungus, virus, prion, or viroid. Microorganisms that cause disease are called pathogens, and the diseases they cause are called infectious diseases. Even when a pathogen is the principal cause of a disease, environmental and hereditary factors often influence the severity of the disease, and whether a particular host individual becomes infected when exposed to the pathogen.

The germ theory was proposed by Girolamo Fracastoro in 1546, but scientific evidence in support of this accumulated slowly and Galen's miasma theory remained dominant among scientists and doctors. A transitional period began in the late 1850s as the work of Louis Pasteur and Robert Koch provided convincing evidence; by 1880, miasma theory was still competing with the germ theory of disease. Eventually, a "golden era" of bacteriology ensued, in which the theory quickly led to the identification of the actual organisms that cause many diseases. Viruses were discovered in the 1890s.

Miasma Theory

A representation of the cholera epidemic of the 19th century depicts the spread of the disease in the form of poisonous air.

The miasma theory of disease transmission held that diseases such as cholera, chlamydia or the Black Death were caused by a *miasma*, a noxious form of "bad air". The theory held that the origin of these epidemic diseases was a miasma, emanating from rotting organic matter. Miasma was considered to be a poisonous vapor or mist filled with particles from decomposed matter (miasmata) that caused illnesses. The miasmatic position was that diseases were the product of environmental factors such as contaminated water, foul air, and poor hygienic conditions. Such infection was not passed between individuals but would affect individuals within the locale that gave rise to such vapors. It was identifiable by its foul smell.

This was the predominant theory of disease transmission before the germ theory of disease took hold in the last decade of the 19th century.

Development

Pre-19th Century

There is some evidence that the Ancient Romans had a germ theory of disease that was later forgotten. Notably, Marcus Terentius Varro wrote, in his 'Rerum rusticarum libri III': "Precautions must also be taken in the neighborhood of swamps [...] because there are bred certain minute creatures which cannot be seen by the eyes, which float in the air and enter the body through the mouth and nose and there cause serious diseases."

Girolamo Fracastoro proposed in 1546 that epidemic diseases are caused by transferable seed-like entities that transmit infection by direct or indirect contact, or even without contact over long distances.

Italian physician Francesco Redi provided early evidence against spontaneous generation. He devised an experiment in 1668 in which he used three jars. He placed a meatloaf and egg in each of the three jars. He had one of the jars open, another one tightly sealed, and the last one covered with gauze. After a few days, he observed that the meatloaf in the open jar was covered by maggots, and the jar covered with gauze had maggots on the surface of the gauze. However, the tightly sealed jar had no maggots inside or outside it. He also noticed that the maggots were found only on surfaces that were accessible by flies. From this he concluded that spontaneous generation is not a plausible theory.

Microorganisms are said to be first directly observed in the 1670s by Anton van Leeuwenhoek, an early pioneer in microbiology. Yet Athanasius Kircher may have done so prior. "When Rome was struck by the bubonic plague in 1656, Kircher spent days on end caring for the sick. Searching for a cure, Kircher observed microorganisms under the microscope and invented the germ theory of disease, which he outlined in his *Scrutinium pestis physico-medicum* (Rome 1658)." Building on Leeuwenhoek's work, physician Nicolas Andry argued in 1700 that microorganisms he called "worms" were responsible for smallpox and other diseases.

In 1720, Richard Bradley theorised that plague and 'all pestilential distempers' were caused by 'poisonous insects', living creatures viewable only with the help of microscopes.

In a 1767 report to the College of Physicians in London, John Zephaniah Holwell mentions the practice of Smallpox vaccinations by Ayurvedic doctors and their explanations of the cause of the disease. According to his report, the Ayurvedic doctors explained to him that disease is caused by 'multitudes of imperceptible animalculae (microorganisms) floating in the atmosphere; that these are the cause of all epidemical diseases'. They further explained to him that they (microorganisms) circulate in and out of the bodies of all animals during respiration and that the severity of disease is proportional to the air charged with the animalculae, and the quantity of them received with the food' He was convinced by the physicians that when the disease of smallpox has run its course either naturally or after inoculations in its weak form, the patients were safe. The major difference between the two being that while natural infections are mostly fatal, the inoculations were merely an inconvenience. According to Henderson and Moss the practice of this method of inoculation has been prevalent in India from just before 1000 A.D.

Agostino Bassi

The Italian Agostino Bassi was the first person to prove that a disease was caused by a microorganism when he conducted a series of experiments between 1808 and 1813, demonstrating that a "vegetable parasite" caused a disease in silkworms known as calcinaccio. This disease was devastating the French silk industry at the time. The "vegetable parasite" is now known to be a fungus pathogenic to insects called *Beauveria bassiana* (named after Bassi).

Ignaz Semmelweis

Ignaz Semmelweis was a Hungarian obstetrician working at the Vienna General Hospital (Allgemeines Krankenhaus) in 1847, when he noticed the dramatically high incidence of death from puerperal fever among women who delivered at the hospital with the help of the doctors and medical students. Births attended by the midwives were relatively safe. Investigating further, Semmelweis made the connection between puerperal fever and examinations of delivering women by doctors, and further realized that these physicians had usually come directly from autopsies. Asserting that puerperal fever was a contagious disease and that matter from autopsies were implicated in its development, Semmelweis made doctors wash their hands with chlorinated lime water before examining pregnant women, thereby reducing mortality from childbirth from 18% to 2.2% at his hospital. Nevertheless, he and his theories were rejected by most of the contemporary medical establishment.

John Snow

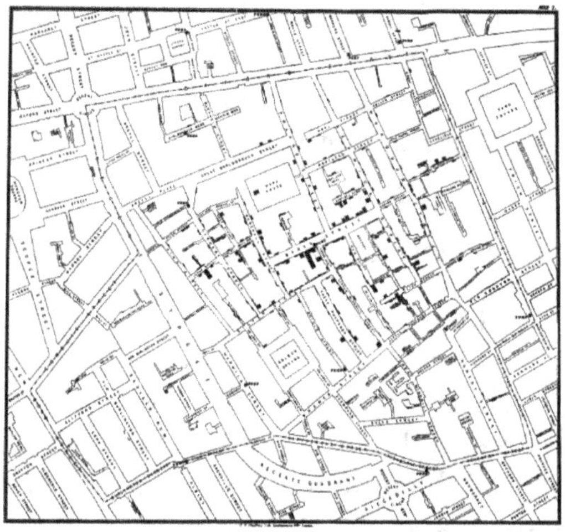

Original map by John Snow showing the clusters of cholera cases in the London epidemic of 1854.

John Snow was a skeptic of the then-dominant miasma theory. Even though the germ theory of disease pioneered by Girolamo Fracastoro had not yet achieved full development or widespread currency, Snow demonstrated a clear understanding of germ theory in his writings. He first published his theory in an 1849 essay *On the Mode of Communication of Cholera,* in which he correctly suggested that the fecal-oral route was the mode of communication, and that the disease replicated itself in the lower intestines. He even proposed in his 1855 edition of the work, that the structure of cholera was that of a cell.

Having rejected effluvia and the poisoning of the blood in the first instance, and being led to the conclusion that the disease is something that acts directly on the alimentary canal, the excretions of the sick at once suggest themselves as containing some material which being accidentally swallowed might attach itself to the mucous membrane of the small intestines, and there multiply itself by appropriation of surrounding matter, in virtue of molecular changes going on within it, or capable of going on, as soon as it is placed in congenial circumstances.

—John Snow (1849)

Snow's 1849 recommendation that water be "filtered and boiled before it is used" is one of the first practical applications of germ theory in the area of public health and is the antecedent to the modern boil water advisory.

In 1855 he published a second edition of his article, documenting his more elaborate investigation of the effect of the water supply in the Soho, London epidemic of 1854.

By talking to local residents, he identified the source of the outbreak as the public water pump on Broad Street (now Broadwick Street). Although Snow's chemical and microscope examination of a water sample from the Broad Street pump did not conclusively prove its danger, his studies of the pattern of the disease were convincing enough to persuade the local council to disable the well pump by removing its handle. This action has been commonly credited as ending the outbreak, but Snow observed that the epidemic may have already been in rapid decline.

Snow later used a dot map to illustrate the cluster of cholera cases around the pump. He also used statistics to illustrate the connection between the quality of the water source and cholera cases. He showed that the Southwark and Vauxhall Waterworks Company was taking water from sewage-polluted sections of the Thames and delivering the water to homes, leading to an increased incidence of cholera. Snow's study was a major event in the history of public health and geography. It is regarded as one of the founding events of the science of epidemiology.

Later, researchers discovered that this public well had been dug only three feet from an old cesspit, which had begun to leak fecal bacteria.The diapers of a baby, who had contracted cholera from another source, had been washed into this cesspit. Its opening was originally under a nearby house, which had been rebuilt farther away after a fire. The city had widened the street and the cesspit was lost. It was common at the time to have a cesspit under most homes. Most families tried to have their raw sewage collected and dumped in the Thames to prevent their cesspit from filling faster than the sewage could decompose into the soil.

After the cholera epidemic had subsided, government officials replaced the handle on the Broad Street pump. They had responded only to the urgent threat posed to the population, and afterward they rejected Snow's theory. To accept his proposal would have meant indirectly accepting the fecal-oral method transmission of disease, which they dismissed as "too depressing."

Louis Pasteur

The more formal experiments on the relationship between germ and disease were conducted by Louis Pasteur between 1860 and 1864. He discovered the pathology of the puerperal fever and the pyogenic vibrio in the blood, and suggested using boric acid to kill these microorganisms before and after confinement.

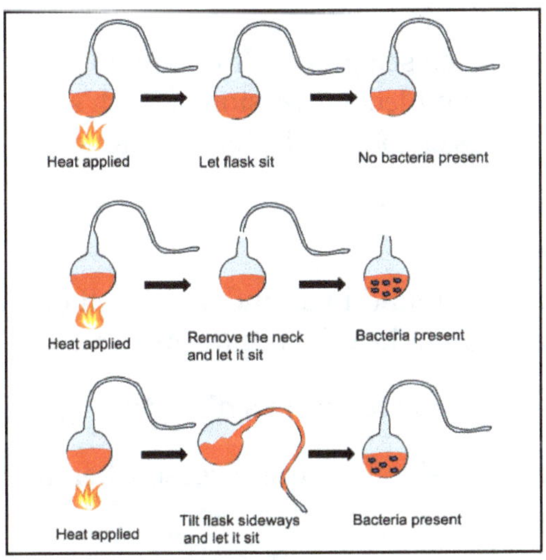

Louis Pasteur's pasteurization experiment illustrates the fact that the spoilage of liquid was caused by particles in the air rather than the air itself. These experiments were importantpieces of evidence supporting the idea of Germ Theory of Disease.

Pasteur further demonstrated between 1860 and 1864 that fermentation and the growth of microorganisms in nutrient broths did not proceed by spontaneous generation. He exposed freshly boiled broth to air in vessels that contained a filter to stop all particles passing through to the growth medium, and even with no filter at all, with air being admitted via a long tortuous tube that would not pass dust particles. Nothing grew in the broths: therefore the living organisms that grew in such broths came from outside, as spores on dust, rather than being generated within the broth.

Pasteur discovered that another serious disease of silkworms, pébrine, was caused by a small microscopic organism now known as Nosema *bombycis* (1870). Pasteur saved the silk industry in France by developing a method to screen silkworms eggs for those that are not infected, a method that is still used today to control this and other silkworm diseases.

Robert Koch

Robert Koch is known for developing four basic criteria (known as Koch's postulates) for demonstrating, in a scientifically sound manner, that a disease is caused by a particular organism. These postulates grew out of his seminal work with anthrax using purified cultures of the pathogen that had been isolated from diseased animals.

Koch's postulates were developed in the 19th century as general guidelines to identify pathogens that could be isolated with the techniques of the day. Even in Koch's time, it was recognized that some infectious agents were clearly responsible for disease even though they did not fulfill all of the postulates. Attempts to rigidly apply Koch's postulates to the diagnosis of viral diseases in the late 19th century, at a time when viruses could not be seen or isolated in culture, may have impeded the early development of the field of virology. Currently, a number of infectious agents are accepted as the cause of disease despite their not fulfilling all of Koch's postulates. Therefore, while Koch's postulates retain historical importance and continue to inform the approach to microbiologic diagnosis, fulfillment of all four postulates is not required to demonstrate causality.

Koch's postulates have also influenced scientists who examine microbial pathogenesis from a molecular point of view. In the 1980s, a molecular version of Koch's postulates was developed to guide the identification of microbial genes encoding virulence factors.

Koch's Postulates:

1. The microorganism must be found in abundance in all organisms suffering from the disease, but should not be found in healthy organisms.
2. The microorganism must be isolated from a diseased organism and grown in pure culture.
3. The cultured microorganism should cause disease when introduced into a healthy organism.
4. The microorganism must be reisolated from the inoculated, diseased experimental host and identified as being identical to the original specific causative agent.

However, Koch abandoned the universalist requirement of the first postulate altogether when he discovered asymptomatic carriers of cholera and, later, of typhoid fever. Asymptomatic or subclinical infection carriers are now known to be a common feature of many infectious diseases, especially viruses such as polio, herpes simplex, HIV, and hepatitis C. As a specific example, all doctors and virologists agree that poliovirus causes paralysis in just a few infected subjects, and the success of the polio vaccine in preventing disease supports the conviction that the poliovirus is the causative agent.

The third postulate specifies "should", not "must", because as Koch himself proved in regard to both tuberculosis and cholera, not all organisms exposed to an infectious agent will acquire the infection. Noninfection may be due to such factors as general health and proper immune functioning; acquired immunity from previous exposure or vaccination; or genetic immunity, as with the resistance to malaria conferred by possessing at least one sickle cell allele.

The second postulate may also be suspended for certain microorganisms or entities that cannot (at the present time) be grown in pure culture, such as prions responsible for Creutzfeldt–Jakob disease. In summary, a body of evidence that satisfies Koch's postulates is sufficient but not necessary to establish causation.

Sanitation

In the 1870s, Joseph Lister was instrumental in developing practical applications of the germ theory of disease with respect to sanitation in medical settings and aseptic surgical techniques—partly through the use of carbolic acid (phenol) as an antiseptic.

Fermentation Theory

Louis Pasteur, Archives Photographiques.

The fermentation theory was studied in depth and brought to light first by Louis Pasteur. This theory states that it is the idea or concept of how fermentation is brought on by microbes and put to the concept of spontaneous generation to rest. Even though this theory is now outdated and has been replaced by the germ theory of disease, for a long time it held true, and Louis was on the forefront of explaining why it seemed organisms appeared out of nothing instead of claiming it was just a spontaneous act of God. Fermentation was a process that has been used for thousands of years, but no one could explain exactly what was happening and why. From Pasteur's discovery of why and how fermentation occurs, the process has been studied intensely and is now a mastered art used in everyday life with processes of making things such as alcoholic beverages, some foods like yogurt or even manufacturing some medications.

Fermentation

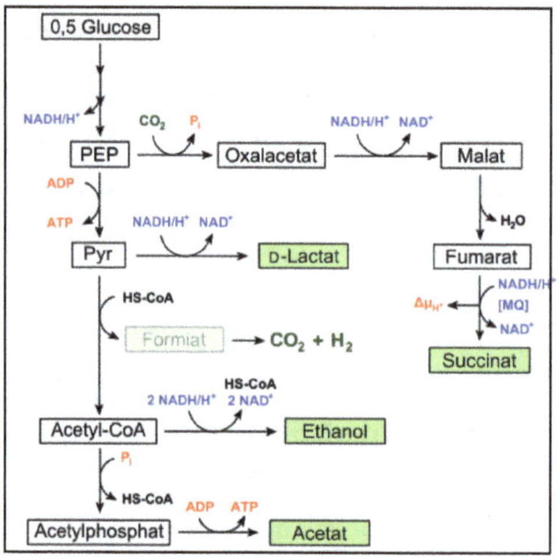

Process of Fermentation.

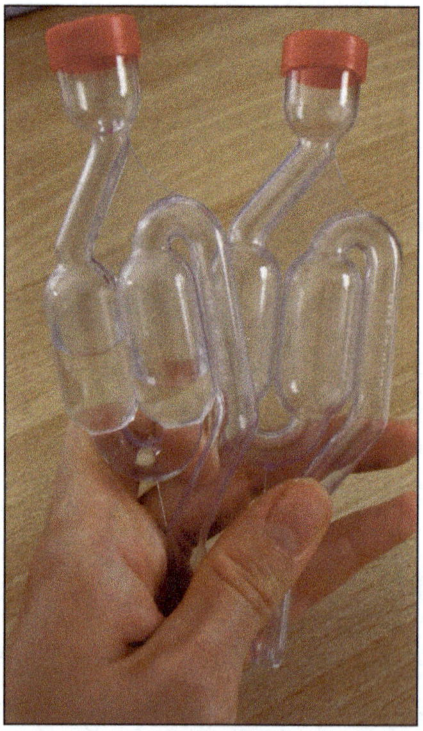

Example of a typical curved neck used today.

Simply put, fermentation is the anaerobic metabolic process that converts sugar into acids, gases, or alcohols. This metabolic process is used in oxygen starved environments. Yeast and many other microbes commonly use this process in order to carry our their anaerobic respiration to survive. Even the human body carries out fermentation processes from time to time. When someone runs a long distance race, lactic acid will build up in their muscles over the course of the race. That lactic acid is the by-product of fermentation taking place in their body, which tries to produce ATP so the body can continue to run since they could not process the oxygen intake fast enough. Although fermentation will give a lower yield of ATP production than aerobic respiration does, it can occur at a much higher rate. Fermentation has been used by humans consciously since around 5000 BCE where there were jars recovered in the Iran Zagros Mountains area in which contained remnants of a microbes similar those present in the process of making grapes into wine.

What the Theory is

Before the 1870s, when Pasteur published his work on this theory, it was believe that microorganisms and even some small animals such as frogs would spontaneously appear, which was coined as spontaneous generation. Spontaneous generation was the explained theory that when elements of the Earth such as clay or mud would mix with water and sunlight in certain amounts, creatures would just appear out of that concoction. A common way that this idea was "proven" over and over again was by taking a piece of raw meat and placing it in open air, which would almost always produce maggots. This idea was accepted and believe to be true before Louis Pasteur shook the Earth with his new ideas that organisms actually came from traceable beginnings. Pasteur demonstrated that fermentation is caused by the growth of microorganisms, and the emergent growth of bacteria in nutrient broths is due to biogenesis rather than spontaneous generation. He exposed boiled broths to air in vessels that contained a filter to prevent all particles from passing through to the growth medium. Yet, when the vessels were open to the air surrounding it, the organisms appeared. It could be concluded that spontaneous generation could be disproven. The organisms did not just appear but were coming from the air, yet we were not able to see them at such a small level. His famous experiment was used with a curved neck placed on top of a beaker. This curved neck was thc kcy to proving his findings because it showed that the germs and microbes had to fall into the broth inside. The curved neck did not allow this to happen.

The fermentation theory of disease is the (now obsolete) concept that many diseases, including the diseases which were "epidemic, endemic and contagious", owe their origin to the presence of a "morbific principle" in the system, acting in a manner analogous to, although not identical with, the process of fermentation. It was rendered obsolete by the germ theory of disease, which led to the new science of bacteriology.

Application

Fermenting a broth in a beaker has become more than just a way to prove that organisms don't just appear out of thin air. Today, the process of fermentation is used for a multitude of everyday applications. Some of those processes include medications which we ingest, beverages we consume, and even food we eat. Currently, companies like Genencor International uses the production of enzymes involved in fermentation to build a revenue of over $400 million a year. Many medications such as antibiotics are produced by the fermentation process. An example is the important drug cortisone, which can be prepared by the fermentation of a plant steroid known as diosgenin.

Burton Union fermentation system, Coors Visitor Centre.

The enzymes used in the reaction are provided by the mold Rhizopus nigricans. Just as it is commonly known, alcohol of all types and brands are also produced by way of fermentation and distillation. Moonshine is a classic example of how this is carried out. Finally, foods such as yogurt are made by fermentation processes as well. Yogurt is a fermented milk product that contains the characteristic bacterial cultures Lactobacillus bulgaricus and Streptococcus thermopiles.

Hologenome Theory of Evolution

The hologenome theory of evolution, also known as the hologenome concept of evolution, recasts the individual animal or plant (and other multicellular organisms) as a community or a "holobiont" - the host plus all of its symbiotic microbes. Consequently, the collective genomes of the holobiont form a "hologenome". Holobionts and hologenomes are structural entities that replace misnomers in the context of host-microbiota symbioses such as superorganism (i.e., an integrated social unit composed of conspecifics), organ, and metagenome. Variation in the hologenome may encode phenotypic plasticity of the holobiont and can be subject to evolutionary changes caused by selection and drift, if portions of the hologenome are transmitted between generations with reasonable fidelity. One of the important outcomes of recasting the individual as a holobiont subject to evolutionary forces is that genetic variation in the hologenome can be brought about by changes in the host genome and also by changes in the microbiome, including new acquisitions of microbes, horizontal gene transfers, and changes in microbial abundance within hosts. Although there is a rich literature on binary host–microbe symbioses, the hologenome concept distinguishes itself by including the vast symbiotic complexity inherent in many multicellular hosts. For recent literature on holobionts and hologenomes published in an open access platform.

The Origin of the Hologenome Theory

In 1991, Lynn Margulis coined the term holobiont in a book chapter entitled Symbiogenesis and

Symbionticism in *Symbiosis as a Source of Evolutionary Innovation: Specation and Morphogenesis* (MIT Press).

In September 1994, Richard Jefferson first introduced the term hologenome at a presentation at Cold Spring Harbor Laboratory during a Symposium "A Decade of PCR", published by Cold Spring Harbor Laboratory Press as a video series. At the CSH Symposium and earlier, the unsettling number and diversity of microbes that were being discovered through the powerful tool of PCR-amplification of 16S ribosomal RNA genes was exciting, but confusing interpretations in diverse studies. A number of speakers referred to microbial contributions to mammalian or plant DNA samples as 'contamination'. In his lecture, Jefferson argued that these were likely not contamination, but rather essential components of the samples that reflected the actual genetic composition of the organism being studied, and that the logic of the organism's performance and capabilities would be embedded only in the hologenome. Observations on the ubiquity of microbes in plant and soil samples as well as laboratory work on molecular genetics of vertebrate-associated microbial enzymes informed this assertion.

In 2008, Eugene Rosenberg and Ilana Zilber-Rosenburg independently derived the term hologenome and developed the hologenome theory of evolution. This theory was originally based on the pair's observations of Vibrio shiloi-mediated bleaching of the coral Oculina patagonica; and since its first introduction, the theory has been promoted as a fusion of Lamarckism and Darwinism and expanded to all of evolution, not just that of corals. The history of the development of the hologenome theory and the logic undergirding its development was the focus of a cover article by Carrie Arnold in New Scientist in January, 2013. The most comprehensive treatment of the theory, including updates by the Rosenbergs on neutrality, pathogenesis and multi-level selection, can be found in their 2013 book.

In 2013, Robert Brucker and Seth Bordenstein re-invigorated the hologenome concept by showing that gut microbiomes, for the first time, are distinguishable between very closely related Nasonia wasp species and contribute to hybrid death. This publication has formed the basis of the recent resurgence and discussion in holobionts and hologenomes, namely by equating interactions between hosts and microbes as part of a conceptual continuum to interactions between genes in the same genome. In 2015, Seth R. Bordenstein and Kevin R. Theis outlined additional history and summarized a conceptual framework that aligns with pre-existing theories in biology. They discuss questions and critiques, show how neutral and selective processes are applicable to the intellectual framework, and discuss future research questions.

Observations from Vertebrate Biology Supporting Hologenome Theory

Multicellular life is made possible by the coordination of physically and temporally distinct processes, most prominently through hormones. Hormones mediate critical activities in vertebrates, including ontogeny, somatic and reproductive physiology, sexual development, performance and behaviour.

Many of these hormones - including most steroids and thyroxines - are secreted, typically as inactivated hormone conjugates to which a glucuronic acid or sulfate has been attached, through the endocrine and apocrine systems into epithelial corridors in which microbiota are widespread and diverse, including gut, urinary tract, lung and skin. There, the inactivated hormones can be re-activated by cleavage of the glucuronide or sulfate residue, allowing the newly re-activated hormone to be reabsorbed. Thus the concentration and bioavailability of many of the hormones is impacted by microbial cleavage of conjugated intermediaries, itself determined by a highly complex and diverse population with redundant enzymatic capabilities. Aspects of this phenomenon of entero-hepatic circulation have been known for decades, and represented in a very large and complex literature, but had previously been viewed as an ancillary effect of detoxification and excretion of metabolites and xenobiotics, including effects on lifetimes of pharmaceuticals, including birth control formulations.

The basic premise is that through the microbial population structure and activity and the diversity of enzyme specificities and kinetic parameters produced by these populations, a spectrum of hormones can be re-activated and resorbed from epithelia, potentially modulating effective time and dose relationships of many vertebrate hormones. The ability to alter and modulate, amplify and suppress, disseminate and recruit new capabilities as microbially-encoded 'traits' means that sampling, sensing and responding to the environment become intrinsic features and emergent capabilities of the holobiont, with mechanisms that can provide rapid, sensitive, nuanced and persistent performance changes.

Studies by Froebe et al. in 1990 indicating that essential mating pheromones, including androstenols, required activation by skin-associated microbial glucuronidases and sulfatases. In the absence of microbial populations in the skin, no detectable aromatic pheromone was released, as the pro-pheromone remained water-soluble and non-volatile. This effectively meant that the microbes in the skin were essential to produce a mating signal.

Observations from Coral Biology Supporting a Hologenome Theory

Subsequent re-articulation describing the hologenome theory by Rosenberg and Zilber-Rosenberg, published 13 years after Jefferson's definition of the theory, was based on their observations of corals, and the coral probiotic hypothesis.

Unbleached and bleached coral.

Coral reefs are the largest structures created by living organisms, and contain abundant and highly complex microbial communities. A coral "head" is a colony of genetically identical polyps, which secrete an exoskeleton near the base. Depending on the species, the exoskeleton may be hard, based on calcium carbonate, or soft and proteinaceous. Over many generations, the colony creates a large skeleton that is characteristic of the species. Diverse forms of life take up residence in a coral colony, including photosynthetic algae such as *Symbiodinium*, as well as a wide range of bacteria including nitrogen fixers, and chitin decomposers, all of which form an important part of coral nutrition. The association between coral and its microbiota is species dependent, and different bacterial populations are found in mucus, skeleton and tissue from the same coral fragment.

Over the past several decades, major declines in coral populations have occurred. Climate change, water pollution and overfishing are three stress factors that have been described as leading to disease susceptibility. Over twenty different coral diseases have been described, but of these, only a handful have had their causative agents isolated and characterized.

Coral bleaching is the most serious of these diseases. In the Mediterranean Sea, the bleaching of *Oculina patagonica* was first described in 1994 and, through a rigorous application of Koch's Postulates, determined to be due to infection by *Vibrio shiloi*. From 1994 to 2002, bacterial bleaching of *O. patagonica* occurred every summer in the eastern Mediterranean. Surprisingly, however, after 2003, *O. patagonica* in the eastern Mediterranean has been resistant to *V. shiloi* infection, although other diseases still cause bleaching.

The surprise stems from the knowledge that corals are long lived, with lifespans on the order of decades, and do not have adaptive immune systems. Their innate immune systems do not produce antibodies, and they should seemingly not be able to respond to new challenges except over evolutionary time scales. Yet multiple researchers have documented variations in bleaching susceptibility that may be termed 'experience-mediated tolerance'. The puzzle of how corals managed to acquire resistance to a specific pathogen led Eugene Rosenberg and Ilana Zilber-Rosenberg to propose the Coral Probiotic Hypothesis. This hypothesis proposes that a dynamic relationship exists between corals and their symbiotic microbial communities. Beneficial mutations can arise and spread among the symbiotic microbes much faster than in the host corals. By altering its microbial composition, the "holobiont" can adapt to changing environmental conditions far more rapidly than by genetic mutation and selection in the host species alone.

Extrapolating the Coral Probiotic Hypothesis to other organisms, including higher plants and animals, led to the Rosenberg's support for and publications around the Hologenome Theory of Evolution.

Hologenome Theory

Definition

The framework of the hologenome theory of evolution is as follows (condensed from Rosenberg *et al.*, 2007):

- "All animals and plants establish symbiotic relationships with microorganisms."
- "Different host species contain different symbiont populations and individuals of the same species can also contain different symbiont populations."

- "The association between a host organism and its microbial community affect both the host and its microbiota."
- "The genetic information encoded by microorganisms can change under environmental demands more rapidly, and by more processes, than the genetic information encoded by the host organism."
- "... the genome of the host can act in consortium with the genomes of the associated symbiotic microorganisms to create a hologenome. This hologenome...can change more rapidly than the host genome alone, thereby conferring greater adaptive potential to the combined holobiont evolution."
- "Each of these points taken together [led Rosenberg *et al.* to propose that] the holobiont with its hologenome should be considered as the unit of natural selection in evolution."

Some authors supplement the above principles with an additional one. If a given holobiont is to be considered a unit of natural selection:

- The hologenome must be heritable from generation to generation.

Ten principles of holobionts and hologenomes were presented in PLOS Biology to clarify and append what these terms and concepts are and are not, to explain how they both support and extend existing theory in the life sciences, and to discuss their potential ramifications for the multifaceted approaches of zoology and botany.

- I. Holobionts and hologenomes are units of biological organization.
- II. Holobionts and hologenomes are not organ systems, superorganisms, or metagenomes.
- III. The hologenome is a comprehensive gene system.
- IV. The hologenome concept reboots elements of Lamarckian evolution.
- V. Hologenomic variation integrates all mechanisms of mutation.
- VI. Hologenomic evolution is most easily understood by equating a gene in the nuclear genome to a microbe in the microbiome.
- VII. The hologenome concept fits squarely into genetics and accommodates multilevel selection theory.
- VIII. The hologenome is shaped by selection and neutrality.
- IX. Hologenomic speciation blends genetics and symbiosis.
- X. Holobionts and their hologenomes do not change the rules of evolutionary biology.

Horizontally Versus Vertically Transmitted Symbionts

Many case studies clearly demonstrate the importance of an organism's associated microbiota to its existence. However, horizontal *versus* vertical transmission of endosymbionts must be distinguished. Endosymbionts whose transmission is predominantly vertical may be considered as contributing to the heritable genetic variation present in a host species.

In the case of colonial organisms such as corals, the microbial associations of the colony persist even though individual members of the colony, reproducing asexually, live and die. Corals also have sexual mode of reproduction, resulting in planktonic larva; it is less clear whether microbial associations persist through this stage of growth. Also, the bacterial community of a colony may change with the seasons.

Many insects maintain heritable obligate symbiosis relationships with bacterial partners. For example, normal development of female wasps of the species *Asobara tabida* is dependent on *Wolbachia* infection. If "cured" of the infection, their ovaries degenerate. Transmission of the infection is vertical through the egg cytoplasm.

In contrast, many obligate symbiosis relationships have been described in the literature where transmission of the symbionts is via horizontal transfer. A well-studied example is the nocturnally feeding squid *Euprymna scolopes*, which camouflages its outline against the moonlit ocean surface by emitting light from its underside with the aid of the symbiotic bacterium *Vibrio fischeri*. The Rosenbergs cite this example within the context of the hologenome theory of evolution. Squid and bacterium maintain a highly co-evolved relationship. The newly hatched squid collects its bacteria from the sea water, and lateral transfer of symbionts between hosts permits faster transfer of beneficial mutations within a host species than are possible with mutations within the host genome.

Primary Versus Secondary Symbionts

Green Vegetable Bug (Nezara viridula) in Fronton, France.

Another traditional distinction between endosymbionts has been between primary and secondary symbionts. Primary endosymbionts reside in specialized host cells that may be organized into larger, organ-like structures (in insects, the bacteriome). Associations between hosts and primary endosymbionts are usually ancient, with an estimated age of tens to hundreds of millions of years. According to endosymbiotic theory, extreme cases of primary endosymbionts include mitochondria, plastids (including chloroplasts), and possibly other organelles of eukaryotic cells. Primary endosymbionts are usually transmitted exclusively vertically, and the relationship is always mutualistic and generally obligate for both partners. Primary endosymbiosis is surprisingly common. An estimated 15% of insect species, for example, harbor this type of endosymbiont. In contrast, secondary endosymbiosis is often facultative, at least from the host point of view, and the associations are less ancient. Secondary endosymbionts do not reside in specialized host tissues, but may dwell in the body cavity dispersed in fat, muscle, or nervous tissue, or may grow within the gut. Transmission may be via vertical, horizontal, or both vertical and horizontal transfer. The relationship between host and secondary endosymbiont is not necessarily beneficial to the host; indeed, the relationship may be parasitic.

The distinction between vertical and horizontal transfer, and between primary and secondary endosymbiosis is not absolute, but follows a continuum, and may be subject to environmental influences. For example, in the stink bug *Nezara viridula*, the vertical transmission rate of symbionts, which females provide to offspring by smearing the eggs with gastric caeca, was 100% at 20 °C, but decreased to 8% at 30 °C. Likewise, in aphids, the vertical transmission of bacteriocytes containing the primary endosymbiont *Buchnera* is drastically reduced at high temperature. In like manner, the distinction between commensal, mutualistic, and parasitic relationships is also not absolute. An example is the relationship between legumes and rhizobial species: N_2 uptake is energetically more costly than the uptake of fixed nitrogen from the soil, so soil N is preferred if not limiting. During the early stages of nodule formation, the plant-rhizobial relationship actually resembles a pathogenesis more than it does a mutualistic association.

Neo-Lamarckism Within a Darwinian Context

Lamarckism, the concept that an organism can pass on characteristics that it acquired during its lifetime to its offspring (also known as inheritance of acquired characteristics or soft inheritance) incorporated two common ideas of its time:

- Use and disuse – individuals lose characteristics they do not require (or use) and develop characteristics that are useful.
- Inheritance of acquired traits – individuals inherit the traits of their ancestors.

Although Lamarckian theory was rejected by the neo-Darwinism of the modern evolutionary synthesis in which evolution occurs through random variations being subject to natural selection, the hologenome theory has aspects that harken back to Lamarckian concepts. In addition to the traditionally recognized modes of variation (*i.e.* sexual recombination, chromosomal rearrangement, mutation), the holobiont allows for two additional mechanisms of variation that are specific to the hologenome theory: (1) changes in the relative population of existing microorganisms (*i.e.* amplification and reduction) and (2) acquisition of novel strains from the environment, which may be passed on to offspring.

Changes in the relative population of existing microorganisms corresponds to Lamarckian "use and disuse", while the ability to acquire novel strains from the environment, which may be passed on to offspring, corresponds to Lamarckian "inheritance of acquired traits". The hologenome theory, therefore, is said by its proponents to incorporate Lamarckian aspects within a Darwinian framework.

Additional Case Studies

Pea aphids extracting sap from the stem and leaves of garden peas.

The pea aphid *Acyrthosiphon pisum* maintains an obligate symbiotic relationship with the bacterium *Buchnera aphidicola,* which is transmitted maternally to the embryos that develop within the mother's ovarioles. Pea aphids live on sap, which is rich in sugars but deficient in amino acids. They rely on their *Buchnera* endosymbiotic population for essential amino acids, supplying in exchange nutrients as well as a protected intracellular environment that allows *Buchnera* to grow and reproduce. The relationship is actually more complicated than mutual nutrition; some strains of *Buchnera* increases host thermotolerance, while other strains do not. Both strains are present in field populations, suggesting that under some conditions, increased heat tolerance is advantageous to the host, while under other conditions, decreased heat tolerance but increased cold tolerance may be advantageous. One can consider the variant *Buchnera* genomes as alleles for the larger hologenome. The association between *Buchnera* and aphids began about 200 million years ago, with host and symbiont co-evolving since that time; in particular, it has been discovered that genome size in various *Buchnera* species has become extremely reduced, in some cases down to 450 kb, which is far smaller even than the 580 kb genome of *Mycoplasma genitalium.*

Development of mating preferences, *i.e.* sexual selection, is considered to be an early event in speciation. In 1989, Dodd reported mating preferences in *Drosophila* that were induced by diet. It has recently been demonstrated that when otherwise identical populations of *Drosophila* were switched in diet between molasses medium and starch medium, that the "molasses flies" preferred to mate with other molasses flies, while the "starch flies" preferred to mate with other starch flies. This mating preference appeared after only one generation and was maintained for at least 37 generations. The origin of these differences were changes in the flies' populations of a particular bacterial symbiont, *Lactobacillus plantarum.* Antibiotic treatment abolished the induced mating preferences. It appears that the symbiotic bacteria changed the levels of cuticular hydrocarbon sex pheromones.

Zilber-Rosenberg and Rosenberg (2008) have tabulated many of the ways in which symbionts are transmitted and their contributions to the fitness of the holobiont, beginning with mitochondria found in all eukaryotes, chloroplast in plants, and then various associations described in specific systems. The microbial contributions to host fitness included provision of specific amino acids, growth at high temperatures, provision of nutritional needs from cellulose, nitrogen metabolism, recognition signals, more efficient food utilization, protection of eggs and embryos against metabolism, camouflage against predators, photosynthesis, breakdown of complex polymers, stimulation of the immune system, angiogenesis, vitamin synthesis, fiber breakdown, fat storage, supply of minerals from the soil, supply of organics, acceleration of mineralization, carbon cycling, and salt tolerance.

Criticism

The hologenome theory is debated. A major criticism by Ainsworth *et al.* has been their claim that *V. shiloi* was misidentified as the causative agent of coral bleaching, and that its presence in bleached *O. patagonica* was simply that of opportunistic colonization.

If this is true, the original observation that led to Rosenberg's later articulation of the theory would be invalid. On the other hand, Ainsworth *et al.* performed their samplings in 2005, two years after the Rosenberg group discovered *O. patagonica* no longer to be susceptible to *V. shiloi* infection; therefore their finding that bacteria are not the primary cause of present-day bleaching in Mediterranean coral *O. patagonica* should not be considered surprising. The rigorous satisfaction of Koch's postulates, as employed in Kushmaro *et al.* (1997), is generally accepted as providing a definitive identification of infectious disease agents.

Baird *et al.* (2009) have questioned basic assumptions made by Reshef *et al.* (2006) in presuming that (1) coral generation times are too slow to adjust to novel stresses over the observed time scales, and that (2) the scale of dispersal of coral larvae is too large to allow for adaptation to local environments. They may simply have underestimated the potential rapidity of conventional means of natural selection. In cases of severe stress, multiple cases have been documented of ecologically significant evolutionary change occurring over a handful of generations. Novel adaptive mechanisms such as switching symbionts might not be necessary for corals to adjust to rapid climate change or novel stressors.

Organisms in symbiotic relationships evolve to accommodate each other, and the symbiotic relationship increases the overall fitness of the participant species. Although the hologenome theory is still being debated, it has gained a significant degree of popularity within the scientific community as a way of explaining rapid adaptive changes that are difficult to accommodate within a traditional Darwinian framework.

Definitions and uses of the words holobiont and hologenome also differ between proponents and skeptics, and the misuse of the terms has led to confusions over what comprises evidence related to the hologenome. Ongoing discourse is attempting to clear this confusion. Theis et al. clarify that "critiquing the hologenome concept is not synonymous with critiquing coevolution, and arguing that an entity is not a primary unit of selection dismisses the fact that the hologenome concept has always embraced multilevel selection."

For instance, Chandler and Turelli (2014) criticize the conclusions of Brucker and Bordenstein (2013), noting that their observations are also consistent with an alternative explanation. Brucker and Bordenstein (2014) responded to these criticisms, claiming they were unfounded because of factual inaccuracies and altered arguments and definitions that were not advanced by Brucker and Bordenstein (2013).

Recently, Forest L Rohwer and colleagues developed a novel statistical test to examine the potential for the hologenome theory of evolution in coral species. They found that coral species do not inherit microbial communities, and are instead colonized by a core group of microbes that associate with a diversity of species. The authors conclude: "Identification of these two symbiont communities supports the holobiont model and calls into question the hologenome theory of evolution." However, other studies in coral adhere to the original and pluralistic definitions of holobionts and hologenomes. David Bourne, Kathleen Morrow and Nicole Webster clarify that "The combined genomes of this coral holobiont form a coral hologenome, and genomic interactions within the hologenome ultimately define the coral phenotype."

Germ Theory Denialism

Germ theory denialism is the belief that germs do not cause infectious disease, and that the germ theory of disease is wrong. It usually involves arguing that Louis Pasteur's model of infectious disease was wrong, and that Antoine Béchamp's was right. One of the first movements to deny the germ theory was the Sanitary Movement, which was nevertheless central in developing America's public health infrastructure. One well-known advocate of this form of denialism is Bill Maher, who has claimed that Pasteur recanted germ theory on his deathbed. Shikha Dalmia, writing in *The Washington Examiner*, referred to Maher as a "germ theory denier" after he made these comments on *Real Time with Bill Maher* on March 4, 2007. However, in response to criticism of his views, Maher said, on the October 16, 2009 episode of his show, that he accepted microorganisms as the cause of some disease, but expressed skepticism about other topics in medicine, such as vaccination. Similarly, the following month, Maher wrote that he "understand[s] germ theory," but that he still thought that "Western medicine ignores too much [sic] the fact that the terrain in which bacteria can thrive is crucial and often controllable, which shouldn't even be controversial." Harriet Hall published an article in *Skeptic* where she describes her experience arguing with germ theory denialists.

An opposite kind of pseudoscience is the belief that germs cause *all* diseases, including ones like cancer, which the preponderance of evidence indicates are not caused by microorganisms. That belief was promoted by Hulda Regehr Clark in her book *The Cure for all Cancers*.

References

- Number 6 in a series of 7 VHS recordings, 'A Decade of PCR: Celebrating 10 Years of Amplification,' released by Cold Spring Harbor Laboratory Press, 1994. ISBN 0-87969-473-4. http://www.cshlpress.com/default.tpl?
- Kushmaro A, Kramarsky-Winter E (2004). "Bacteria as a source of coral nutrition". In Rosenberg E, Loya Y. Coral Health and Disease. Berlin Heidelberg: Springer-Verlag. pp. 231–241. ISBN 3-540-20772-4
- Maher, Bill (15 November 2009). "Vaccination: A Conversation Worth Having". Huffington Post. Retrieved 22, February 2020
- Rosenberg, Eugene; Zilber-Rosenberg, Ilana (2016-05-04). "Microbes Drive Evolution of Animals and Plants: the Hologenome Concept". mBio. 7 (2): e01395–15. doi:10.1128/mBio.01395-15. ISSN 2150-7511. PMC 4817260 . PMID 27034283

- Dalmia, Shikha (10 May 2013). “Shikha Dalmia: ‘Scientific’ liberals should accept results of science”. The Washington Examiner. Retrieved 15, May 2020
- Feldhaar H (2011). “Bacterial symbionts as mediators of ecologically important traits of insect hosts”. Ecological entomology. 5 (5): 533–543. doi:10.1111/j.1365-2311.2011.01318.x

Chapter 6

Microbial Analysis: Methods and Techniques

There is no one singular way to undertake the study of microorganisms. Microbiology has many tried and tested methods and techniques which are employed for research. Microbiological culture, bacteriological water analysis, impedance, streaking, thermal shift assay, ATP test, phenotypic testing of mycobacteria, etc. are some microbiological methods and techniques covered in this chapter.

Microbiological Culture

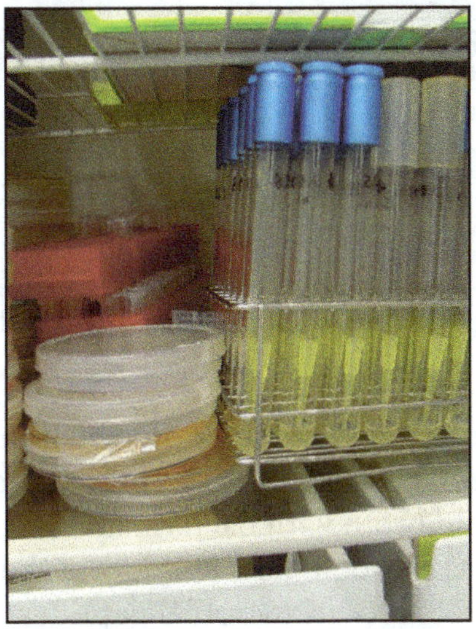

Solid and liquid microbial cultures.

A microbiological culture, or microbial culture, is a method of multiplying microbial organisms by letting them reproduce in predetermined culture media under controlled laboratory conditions. Microbial cultures are used to determine the type of organism, its abundance in the sample being tested, or both. It is one of the primary diagnostic methods of microbiology and used as a tool to determine the cause of infectious disease by letting the agent multiply in a predetermined medium. For example, a throat culture is taken by scraping the lining of tissue in the back of the throat and blotting the sample into a medium to be able to screen for harmful microorganisms, such as *Streptococcus pyogenes*, the causative agent of strep throat. Furthermore, the term culture is more generally used informally to refer to "selectively growing" a specific kind of microorganism in the lab.

Microbial cultures are foundational and basic diagnostic methods used extensively as a research tool in molecular biology. It is often essential to isolate a pure culture of microorganisms. A pure (or *axenic*) culture is a population of cells or multicellular organisms growing in the absence of other species or types. A pure culture may originate from a single cell or single organism, in which case the cells are genetic clones of one another.

For the purpose of gelling the microbial culture, the medium of agarose gel (agar) is used. Agar is a gelatinous substance derived from seaweed. A cheap substitute for agar is guar gum, which can be used for the isolation and maintenance of thermophiles.

Bacterial Culture

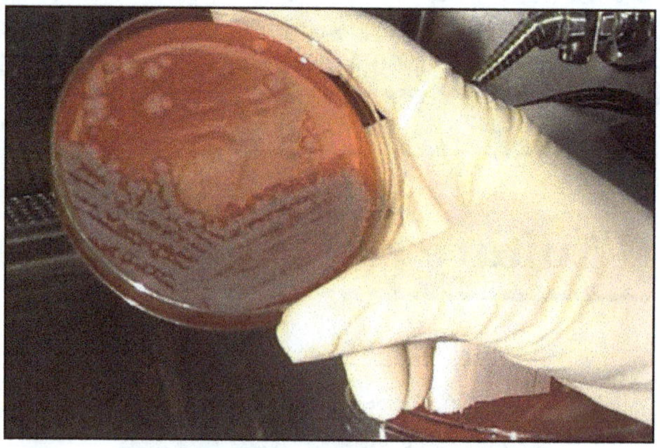

A culture of *Bacillus anthracis*.

Microbiological cultures can be grown in petri dishes of differing sizes that have a thin layer of agar-based growth medium. Once the growth medium in the petri dish is inoculated with the desired bacteria, the plates are incubated at the best temperature for the growing of the selected bacteria (for example, usually at 37 degrees Celsius for cultures from humans or animals, or lower for environmental cultures).

Another method of bacterial culture is liquid culture, in which the desired bacteria are suspended in liquid broth, a nutrient medium. These are ideal for preparation of an antimicrobial assay. The experimenter would inoculate liquid broth with bacteria and let it grow overnight (they may use a shaker for uniform growth). Then they would take aliquots of the sample to test for the antimicrobial activity of a specific drug or protein (antimicrobial peptides).

Liquid cultures of cyanobacterium *Synechococcus* PCC 7002.

As an alternative, the microbiologist may decide to use static liquid cultures. These cultures are not shaken and they provide the microbes with an oxygen gradient.

Culture Collections

Microbial culture collections focus on the acquisition, authentication, production, preservation, catalogueing and distribution of viable cultures of standard reference microorganisms, cell lines and other materials for research in microbial systematics. Culture collection are also repositories of type strains.

Major national culture collections.		
Collection Acronym	**Name**	**Location**
ATCC	American Type Culture Collection	Manassas, Virginia
NCTC	National Collection of Type Cultures	Public Health England, London, United Kingdom
BCCM	Belgium Coordinated Collection of Microorganism	Ghent, Belgium
CIP	Collection d'Institut Pasteur	Paris, France
DSMZ	Deutsche Sammlung von Mikroorganismen und Zellkulturen	Braunschweig, Germany
JCM	Japan Collection of Microorganisms	Tsukuba, Ibaraki, Japan
NCCB	Netherlands Culture Collection of Bacteria	Utrecht, Netherlands
NCIMB	National Collection of Industrial, Food and Marine Bacteria	Aberdeen, Scotland
STCC	Spanish Type Culture Collection, Valencia University	Valencia, Spain

Virus and Phage Culture

Virus or phage cultures require host cells in which the virus or phage multiply. For bacteriophages, cultures are grown by infecting bacterial cells. The phage can then be isolated from the resulting plaques in a lawn of bacteria on a plate. Virus cultures are obtained from their appropriate eukaryotic host cells.

Eukaryotic Cell Culture

Isolation of Pure Cultures

For single-celled eukaryotes, such as yeast, the isolation of pure cultures uses the same techniques as for bacterial cultures. Pure cultures of multicellular organisms are often more easily isolated by simply picking out a single individual to initiate a culture. This is a useful technique for pure culture of fungi, multicellular algae, and small metazoa, for example.

Developing pure culture techniques is crucial to the observation of the specimen in question. The most common method to isolate individual cells and produce a pure culture is to prepare a streak plate. The streak plate method is a way to physically separate the microbial population, and is done by spreading the inoculate back and forth with an inoculating loop over the solid agar plate. Upon incubation, colonies will arise and single cells will have been isolated from the biomass.

Bacteriological Water Analysis

Bacteriological water analysis is a method of analysing water to estimate the numbers of bacteria present and, if needed, to find out what sort of bacteria they are. It represents one aspect of water quality. It is a microbiological analytical procedure which uses samples of water and from these samples determines the concentration of bacteria. It is then possible to draw inferences about the suitability of the water for use from these concentrations. This process is used, for example, to routinely confirm that water is safe for human consumption or that bathing and recreational waters are safe to use.

The interpretation and the action trigger levels for different waters vary depending on the use made of the water. Very stringent levels applying to drinking water whilst more relaxed levels apply to marine bathing waters, where much lower volumes of water are expected to be ingested by users.

Approach

The common feature of all these routine screening procedures is that the primary analysis is for indicator organisms rather than the pathogens that might cause concern. Indicator organisms are bacteria such as non-specific coliforms, *Escherichia coli* and *Pseudomonas aeruginosa* that are very commonly found in the human or animal gut and which, if detected, may suggest the presence of sewage. Indicator organisms are used because even when a person is infected with a more pathogenic bacteria, they will still be excreting many millions times more indicator organisms than pathogens. It is therefore reasonable to surmise that if indicator organism levels are low, then pathogen levels will be very much lower or absent. Judgements as to suitability of water for use are based on very extensive precedents and relate to the probability of any sample population of bacteria being able to be infective at a reasonable statistical level of confidence.

Analysis is usually performed using culture, biochemical and sometimes optical methods. When indicator organisms levels exceed pre-set triggers, specific analysis for pathogens may then be undertaken and these can be quickly detected (where suspected) using specific culture methods or molecular biology.

Methodologies

Because the analysis is always based on a very small sample taken from a very large volume of water, all methods rely on statistical principles.

Multiple Tube Method

One of the oldest methods is called the multiple tube method. In this method a measured sub-sample (perhaps 10 ml) is diluted with 100 ml of sterile growth medium and an aliquot of 10 ml is then decanted into each of ten tubes. The remaining 10 ml is then diluted again and the process repeated. At the end of 5 dilutions this produces 50 tubes covering the dilution range of 1:10 through to 1:10000.

The tubes are then incubated at a pre-set temperature for a specified time and at the end of the process the number of tubes with growth in is counted for each dilution. Statistical tables are then used to derive the concentration of organisms in the original sample. This method can be enhanced by using indicator medium which changes colour when acid forming species are present and by including a tiny inverted tube called a Durham tube in each sample tube. The Durham inverted tube catches any gas produced. The production of gas at 37 degrees Celsius is a strong indication of the presence of *Escherichia coli.*

ATP Testing

An ATP test is the process of rapidly measuring active microorganisms in water through detection adenosine triphosphate (ATP). ATP is a molecule found only in and around living cells, and as such it gives a direct measure of biological concentration and health. ATP is quantified by measuring the light produced through its reaction with the naturally-occurring enzyme firefly luciferase using a luminometer. The amount of light produced is directly proportional to the amount of biological energy present in the sample.

Second generation ATP tests are specifically designed for water, wastewater and industrial applications where, for the most part, samples contain a variety of components that can interfere with the ATP assay.

Plate Count

The plate count method relies on bacteria growing a colony on a nutrient medium so that the colony becomes visible to the naked eye and the number of colonies on a plate can be counted. To be effective, the dilution of the original sample must be arranged so that on average between 30 and 300 colonies of the target bacterium are grown. Fewer than 30 colonies makes the interpretation statistically unsound whilst greater than 300 colonies often results in overlapping colonies and imprecision in the count. To ensure that an appropriate number of colonies will be generated several dilutions are normally cultured. This approach is widely utilised for the evaluation of the effectiveness of water treatment by the inactivation of representative microbial contaminants such as "E. coli" following ASTM D5465.

The laboratory procedure involves making serial dilutions of the sample (1:10, 1:100, 1:1000, etc.) in sterile water and cultivating these on nutrient agar in a dish that is sealed and incubated. Typical media include plate count agar for a general count or MacConkey agar to count Gram-negative bacteria such as *E. coli.* Typically one set of plates is incubated at 22°C and for 24 hours and a second set at 37°C for 24 hours. The composition of the nutrient usually includes reagents that resist the growth of non-target organisms and make the target organism easily identified, often by a colour change in the medium. Some recent methods include a fluorescent agent so that counting of the colonies can be automated. At the end of the incubation period the colonies are counted by eye, a procedure that takes a few moments and does not require a microscope as the colonies are typically a few millimetres across.

Membrane Filtration

Most modern laboratories use a refinement of total plate count in which serial dilutions of the sample are vacuum filtered through purpose made membrane filters and these filters are themselves laid on nutrient medium within sealed plates. The methodology is otherwise similar to conventional total plate counts. Membranes have a printed millimetre grid printed on and can be reliably used to count the number of colonies under a binocular microscope.

Pour Plate Method

When the analysis is looking for bacterial species that grow poorly in air, the initial analysis is done by mixing serial dilutions of the sample in liquid nutrient agar which is then poured into bottles which are then sealed and laid on their sides to produce a sloping agar surface. Colonies that develop in the body of the medium can be counted by eye after incubation.

The total number of colonies is referred to as the Total Viable Count (TVC). The unit of measurement is cfu/ml (or colony forming units per millilitre) and relates to the original sample. Calculation of this is a multiple of the counted number of colonies multiplied by the dilution used.

Pathogen Analysis

When samples show elevated levels of indicator bacteria, further analysis is often undertaken to look for specific pathogenic bacteria. Species commonly investigated in the temperate zone include *Salmonella typhi* and *Salmonella Typhimurium*. Depending on the likely source of contamination investigation may also extend to organisms such as *Cryptosporidium spp*. In tropical areas analysis of *Vibrio cholerae* is also routinely undertaken.

Types of Nutrient Media Used in Analysis

MacConkey agar is culture medium designed to grow Gram-negative bacteria and stain them for lactose fermentation. It contains bile salts (to inhibit most Gram-positive bacteria), crystal violet dye (which also inhibits certain Gram-positive bacteria), neutral red dye (which stains microbes fermenting lactose), lactose and peptone. Alfred Theodore MacConkey developed it while working as a bacteriologist for the Royal Commission on Sewage Disposal in the United Kingdom.

Endo agar contains peptone, lactose, dipotassium phosphate, agar, sodium sulfite, basic fuchsin and was originally developed for the isolation of *Salmonella typhi*, but is now commonly used in water analysis. As in MacConkey agar, coliform organisms ferment the lactose, and the colonies become red. Non-lactose-fermenting organisms produce clear, colourless colonies against the faint pink background of the medium.

mFC medium is used in membrane filtration and contains selective and differential agents. These include Rosolic acid to inhibit bacterial growth in general, except for faecal coliforms, Bile salts inhibit non-enteric bacteria and Aniline blue indicates the ability of faecal coliforms to ferment lactose to acid that causes a pH change in the medium.

TYEA medium contains tryptone, yeast extract, common salt and L-arabinose per liter of glass distilled water and is a non selective medium usually cultivated at two temperatures (22 and 36°C) to determine a general level of contamination (a.k.a. colony count).

Impedance Microbiology

Impedance microbiology is a rapid microbiological technique used to measure the microbial concentration (mainly bacteria but also yeasts) of a sample by monitoring the electrical parameters of the growth medium. The ability of microbial metabolism to change the electrical conductivity of the growth medium was discovered by Stewart and further studied by other scientists such as Oker-Blom, Parson and Allison in the first half of 20th century. However, it was only in the late 1970s that, thanks to computer-controlled systems used to monitor impedance, the technique showed its full potential, as discussed in the works of Fistenberg-Eden & Eden, Ur & Brown and Cady.

Principle of Operation

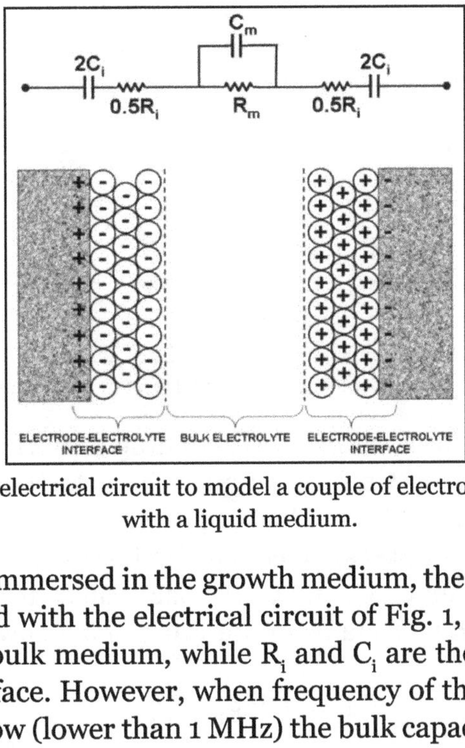

Figure : Equivalent electrical circuit to model a couple of electrodes in direct contact with a liquid medium.

When a pair of electrodes are immersed in the growth medium, the system composed of electrodes and electrolyte can be modeled with the electrical circuit of Fig. 1, where R_m and C_m are the resistance and capacitance of the bulk medium, while R_i and C_i are the resistance and capacitance of the electrode-electrolyte interface. However, when frequency of the sinusoidal test signal applied to the electrodes is relatively low (lower than 1 MHz) the bulk capacitance C_m can be neglected and the system can be modeled with a simpler circuit consisting only of a resistance R_s and a capacitance C_s in series. The resistance R_s accounts for the electrical conductivity of the bulk medium while the capacitance C_s is due to the capacitive double-layer at the electrode-electrolyte interface. During the growth phase, bacterial metabolism transforms uncharged or weakly charged compounds of the bulk medium in highly charged compounds that change the electrical properties of the medium. This results in a decrease of resistance R_s and an increase of capacitance C_s.

Impedance Microbiology technique works this way:

1. The sample with the initial unknown bacterial concentration (C_0) is placed at a temperature favoring bacterial growth (in the range 37 to 42 °C if mesophilic microbial population is the target) and the electrical parameters R_s and C_s are measured at regular time intervals of few minutes by means of a couple of electrodes in direct contact with the sample.

2. Until the bacterial concentration is lower than a critical threshold C_{TH} the electrical parameters R_s and C_s remain essentially constant (at their baseline values). C_{TH} depends on various parameters such as electrode geometry, bacterial strain, chemical composition of the growth medium etc., but it is always in the range 10^6 to 10^7 cfu/ml.
3. When the bacterial concentration increases over C_{TH}, the electrical parameters deviate from their baseline values (generally in the case of bacteria there is a decrease of R_s and an increase of C_s, the opposite happens in the case of yeasts).
4. The time needed for the electrical parameters R_s and C_s to deviate from their baseline value is referred as Detect Time (DT) and is the parameter used to estimate the initial unknown bacterial concentration C_0.

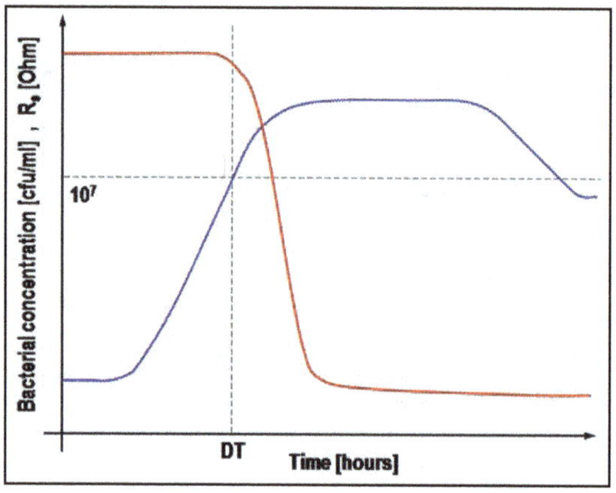

Figure: R_s curve and bacterial concentration curve as function of time.

In Fig. a typical curve for R_s as well as the corresponding bacterial concentration are plotted vs. time. Fig. shows typical R_s curves vs time for samples characterized by different bacterial concentration. Since DT is the time needed for the bacterial concentration to grow from the initial value C_0 to C_{TH}, highly contaminated samples are characterized by lower values of DT than samples with low bacterial concentration. Given C_1, C_2 and C_3 the bacterial concentration of three samples with $C_1 > C_2 > C_3$, it is $DT_1 < DT_2 < DT_3$. Data from literature show how DT is a linear function of the logarithm of C_0:

$$DT = A \cdot \log_{10}(C_0) + B$$

where the parameters A and B are dependent on the particular type of samples under test, the bacterial strains, the type of enriching medium used and so on. These parameters can be calculated by calibrating the system using a set of samples whose bacterial concentration is known and calculating the linear regression line that will be used to estimate the bacterial concentration from the measured DT.

Impedance microbiology has different advantages on the standard Plate Count technique to measure bacterial concentration:

1. It is characterized by faster response time. In the case of mesophilic bacteria, the response time range from 2 – 3 hours for highly contaminated samples (10^5 - 10^6 cfu/ml) to over 10 hours for samples with very low bacterial concentration (less than 10 cfu/ml). As a comparison, for the same bacterial strains the Plate Count technique is characterized by response times from 48 to 72 hours.

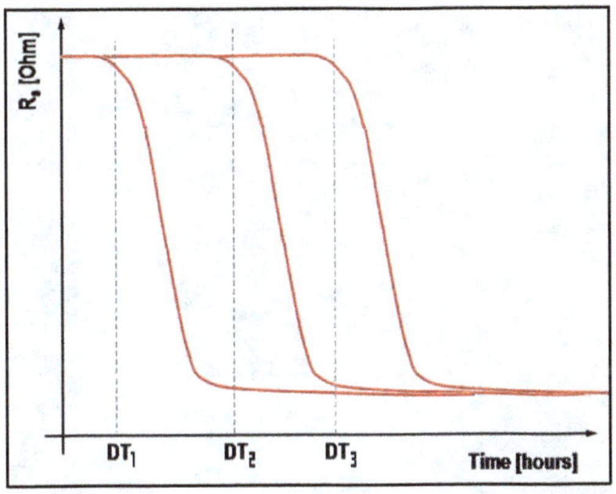

Figure 3: R_s curves for samples featuring different bacterial concentration as function of time.

2. Impedance microbiology is a method that can be easily automated, i.e. implemented as part of an industrial machine or realized as an embedded portable sensor, while Plate Count is a manual method that needs to be carried out in a laboratory by long trained personnel.

Instrumentation

Over the past decades different instruments (either laboratory built or commercially available) to measure bacterial concentration using impedance microbiology have been built. One of the best selling and well accepted instruments in the industry is the Bactometer by Biomerieux. The original instrument of 1984 features a multi-incubator system capable of monitoring up to 512 samples simultaneously with the ability to set 8 different incubation temperatures. Other instruments with performance comparable to the Bactometer are Malthus by Malthus Instruments Ltd (Bury, UK), RABIT by Don Whitley Scientific (Shipley, UK) and Bac Trac by Sy-Lab (Purkensdorf, Austria). A portable embedded system for microbial concentration measurement in liquid and semi-liquid media using impedance microbiology has been recently proposed. The system is composed of a thermoregulated incubation chamber where the sample under test is stored and a controller for thermoregulation and impedance measurements.

Applications

Impedance microbiology has been extensively used in the past decades to measure the concentration of bacteria and yeasts in different type of samples, mainly for quality assurance in the food industry. Some applications are, the determination of the shelf life of pasteurized milk and the measure of total bacterial concentration in raw-milk, frozen vegetables, grain products, meat products and beer. The technique has been also used in environmental monitoring to detect the coliform concentration in water samples, in the pharmaceutical industry to test the efficiency of novel antibacterial agents and the testing of final products.

Streaking

In microbiology, streaking is a technique used to isolate a pure strain from a single species of microorganism, often bacteria. Samples can then be taken from the resulting colonies and a microbiological culture can be grown on a new plate so that the organism can be identified, studied, or tested.

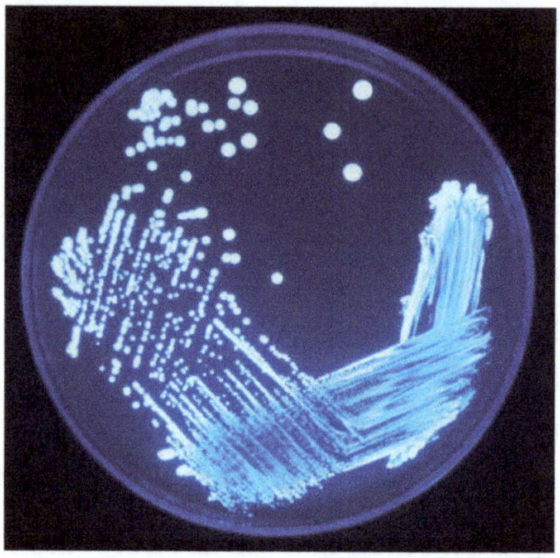

A plate which has been streaked showing the colonies thinning as the streaking moves clockwise.

The modern streak plate method has progressed from the efforts of Robert Koch and other microbiologists to obtain microbiological cultures of bacteria in order to study them. The dilution or isolation by streaking method was first developed by Loeffler and Gaffky in Koch's laboratory, which involves the dilution of bacteria by systematically streaking them over the exterior of the agar in a petri dish to obtain isolated colonies which will then grow into quantity of cells, or isolated colonies. If the agar surface grows microorganisms which are all genetically same, the culture is then considered as a microbiological culture.

Technique

Streaking is rapid and ideally a simple process of isolation dilution. The technique is done by diluting a comparatively large concentration of bacteria to a smaller concentration. The decrease of bacteria should show that colonies are sufficiently spread apart to effect the separation of the different types of microbes. Streaking is done using a sterile tool, such as a cotton swab or commonly an inoculation loop. Aseptic techniques are used to maintain microbiological cultures and to prevent contamination of the growth medium.There are many different types of methods used to streak a plate. Picking a technique is a matter of individual preference and can also depend on how large the number of microbes the sample contains.

The three-phase streaking pattern, known as the T-Streak, is recommended for beginners.The streaking is done using a sterile tool, such as a cotton swab or commonly an inoculation loop. The inoculation loop is first sterilized by passing it through a flame. When the loop is cool, it is dipped into an inoculum such as a broth or patient specimen containing many species of bacteria. The inoculation loop is then dragged across the surface of the agar back and forth in a zigzag motion until approximately 30% of the plate has been covered. The loop then is re-sterilized and the plate is turned 90 degrees. Starting in the previously streaked section, the loop is dragged through it two to three times continuing the zigzag pattern. The procedure is then repeated once more being cautious to not touch the previously streaked sectors. Each time the loop gathers fewer and fewer bacteria until it gathers just single bacterial cells that can grow into a colony.The plate should show the heaviest growth in the first section. The second section will have less growth and a few isolated colonies, while the final section will have the least amount of growth and many isolated colonies.

Growth Medium

The sample is spread across one quadrant of a petri dish containing a growth medium. Bacteria need different nutrients to grow. This includes water, a source of energy, sources of carbon, sulfur, nitrogen, phosphorus, certain minerals, and other vitamins and growth factors. A very common type of media used in microbiology labs is known as agar, a gelatinous substance derived from seaweed. The nutrient agar has a lot of ingredients with unknown amounts of nutrients in them. On one hand, this can be a very selective media to use because as mentioned bacteria are particular. If there is a certain nutrient in the media the bacteria could most certainly not grow and could die. On the other hand, this media is very complex. Complex media is important because it allows for a wide range of microbial growth. The bacteria growth can be supported by this media greatly due in part to the numerous amounts of nutrients. Choice of which growth medium is used depends on which microorganism is being cultured, or selected for.

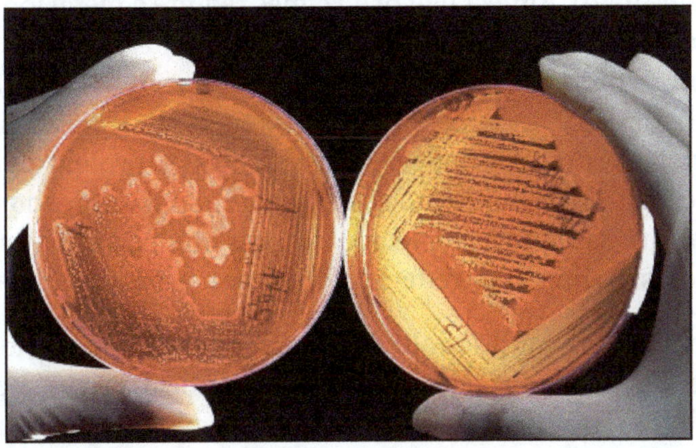

Different labs have different standards as to the direction and style of the streaking.

Incubation

Dependent on the strain, the plate may then be incubated, usually for 24 to 36 hours, to allow the bacteria to reproduce. At the end of incubation there should be enough bacteria to form visible colonies in the areas touched by the inoculation loop. From these mixed colonies, single bacterial or fungal species can be identified based on their morphological (size/shape/colour) differences, and then sub-cultured to a new media plate to yield a pure culture for further analysis.

Automated equipment is used at industrial level for streak plating the solid media in order to achieve better sterilization and consistency of streaking and for reliably faster work. While streaking manually it is important to avoid scratching the solid medium as subsequent streak lines will be damaged and non-uniform deposition of inoculum at damaged sites on the medium yield clustered growth of microbes which may extend into nearby streak lines.

Importance

Bacteria exist in water, soil and food, on skin, and intestinal tract normal flora. The assortment of microbes that exist in the environment and on human bodies is enormous. The human body has billions of bacteria which creates the normal flora fighting against the invading pathogens. Bacteria frequently occur in mixed populations. It is very rare to find a single occurring species of bacteria. To be able to study the cultural, morphological, and physiological characteristics of an individual species, it is vital that the bacteria be divided from the other species that generally originate in the environment. This is important in determining a bacterium in a clinical sample. When the bacteria is streaked and isolated, the causative agent of a bacterial disease can be identified.

Thermal Shift Assay

A thermal shift assay quantifies the change in thermal denaturation temperature of a protein under varying conditions. The differing conditions that can be examined are very diverse, e.g. pH, salts, additives, drugs, drug leads, oxidation/reduction, or mutations. The binding of low molecular weight ligands can increase the thermal stability of a protein, as described by Koshland (1958) and Linderstrom-Lang and Schellman (1959). Almost half of enzymes require a metal ion co-factor. Thermostable proteins are often more useful than their non-thermostable counterparts, e.g. DNA polymerase in the polymerase chain reaction, so protein engineering often includes adding mutations to increase thermal stability. Protein crystallisation is more successful for proteins with a higher melting point and adding buffer components that stabilise proteins improve the likelihood of protein crystals forming. If examining pH then the possible effects of the buffer molecule on thermal stability should be taken into account along with the fact that pKa of each buffer molecule changes uniquely with temperature. Additionally, any time a charged species is examined the effects of the counterion should be accounted for.

Thermal stability of proteins has traditionally been investigated using biochemical assays, circular dichroism, or differential scanning calorimetry. Biochemical assays require a catalytic activity of the protein in question as well as a specific assay. Circular dichroism and differential scanning calorimetry both consume large amounts of protein and are low-throughput methods. The thermofluor assay was the first high-throughput thermal shift assay and its utility and limitations has spurred the invention of a plethora of alternate methods. Each method has its strengths and weaknesses but they all struggle with intrinsically disordered proteins without any clearly defined tertiary structure as the essence of a thermal shift assay is measuring the temperature at which a protein goes from well-defined structure to disorder.

Thermofluor

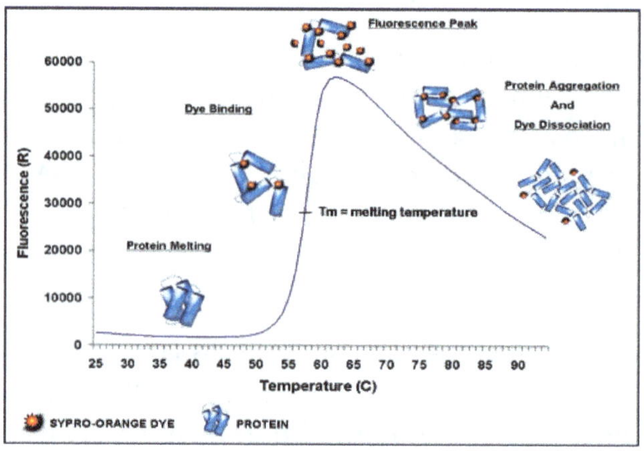

Diagram.

The technique was first described by Semisotnov et al. (1991) using 1,8-ANS and quartz cuvettes. Millennium Pharmaceuticals were the first to describe a high-throughput version using a plate reader and Wyeth Research published a variation of the method with SYPRO Orange instead of 1,8-ANS. SYPRO Orange has an excitation/emission wavelength profile compatible with qPCR machines which are almost ubiquitous in institutions that perform molecular biology research. The name Differential Scanning Fluorimetry (DSF) was introduced later but thermofluor is preferable as thermofluor is no longer trademarked and differential scanning fluorimetry is easily confused with differential scanning calorimetry.

SYPRO Orange binds nonspecifically to hydrophobic surfaces, and water strongly quenches its fluorescence. When the protein unfolds, the exposed hydrophobic surfaces bind the dye, resulting in an increase in fluorescence by excluding water. Detergent micelles will also bind the dye and increase background noise dramatically - this effect is lessened by switching to the dye ANS but that requires UV excitation. The stability curve and its midpoint value (melting temperature, T_m also known as the temperature of hydrophobic exposure, T_h) are obtained by gradually increasing the temperature to unfold the protein and measuring the fluorescence at each point. Curves are measured for protein only and protein + ligand, and ΔTm is calculated. The method might not work very well for protein-protein interactions if one of the interaction partners contains large hydrophobic patches as it is highly non-trivial to dissect prevention of aggregation, stabilisation of a native folds and steric hindrance of dye access to hydrophobic sites. In addition, partly aggregated protein can also limit the relative fluorescence increase upon heating; in extreme cases there will be no fluorescence increase at all because all protein is already in aggregates before heating. Knowing this effect can be very useful as a high relative fluorescence increase suggests a significant fraction of folded protein in the starting material.

This assay allows high-throughput screening of ligands to the target protein and it is widely used in the early stages of drug discovery in the pharmaceutical industry, structural genomics efforts, and high-throughput protein engineering.

A Typical Assay

1. Materials: A fluorometer equipped with temperature control or similar instrumentation (RT-PCR machines); suitable fluorescent dye; a suitable assay plate, such as 96 well RT-PCR plate.

2. Compound solutions: Test ligands are prepared at a 50- to 100-fold concentrated solution, generally in the 10-100 mM range. For titration, a typical experimental protocol employs a set of 12 well, comprising 11 different concentrations of a test compound with a single negative control well.

3. Protein solution: Typically, target protein is diluted from a concentrated stock to a working concentration of ~0.5-5 μM protein with dye into a suitable assay buffer. The exact concentrations of protein and dye are defined by experimental assay development studies.

4. Centrifugation and oil dispense: A brief centrifugation (~1000 × g-force, 1 min) of the assay plate to mix compounds into the protein solution, 1-2 μl silicone oil to prevent the

evaporation during heating is overlaid onto the solution (some systems use plastic seals instead), followed by an additional centrifugation step (~1000 × g-force, 1 min).

5. Instrumental set up: A typical temperature ramp rates range from 0.1-10 °C/min but generally in the range of 1 °C/min. The fluorescence in each well is measured at regular intervals, 0.2-1 °C/image, over a temperature range spanning the typical protein unfolding temperatures of 25-95 °C.

CPM, Thiol-specific Dyes

Alexandrov et al. (2008) published a variation on the thermofluor assay where SYPRO Orange was replaced by N-[4-(7-diethylamino-4-methyl-3-coumarinyl)phenyl]maleimide (CPM), a compound that only fluoresces after reacting with a nucleophile. CPM has a high preference for thiols over other typical biological nucleophiles and therefore will react with cysteine side chains before others. Cysteines are typically buried in the interior of a folded protein as they are hydrophobic. When a protein denatures cysteine thiols become available and a fluorescent signal can be read from reacted CPM. The exitation and emission wavelengths for reacted CPM are 387 nm/ 463 nm so a fluorescence plate reader or a qPCR machine with specialised filters is required. Alexandrov et al. used the technique successfully on the membrane proteins Apelin GPCR and FAAH as well as β-lactoglobin which fibrillates on heating rather than going to a molten globule.

DCVJ, Rigidity Sensitive Dyes

4-(dicyanovinyl)julolidine (DCVJ) is a molecular rotor probe with fluorescence that is strongly dependent on the rigidity of its environment. When protein denatures DCVJ increases in fluorescence. It has been reported to work with 40 mg/ml of antibody.

Intrinsic Tryptophan Fluorescence Lifetime

The lifetime of tryptophan fluorescence differs between folded and unfolded protein and by measuring the lifetime of UV-excited fluorescence at temperature intervals you can get a measurement of the melting temperature of your protein.

A prominent advantage is that no reporter dyes are required to be added as tryptophan is an intrinsic part of the protein. This can also be a disadvantage as not all proteins contain tryptophan. Intrinsic fluorescence lifetime works with membrane proteins and detergent micelles but a powerful UV fluorescer in the buffer could drown out the signal and few articles are published using the technique for thermal shift assays.

Intrinsic Tryptophan Fluorescence Wavelength

The excitation and emission wavelengths of tryptophan are dependent on the immediate environment and therefore differs between folded and unfolded protein, just as the fluorescence lifetime. Currently there are at least two machines on the market that can read this shift in wavelength in a high-throughput manner while heating the samples.

The advantages and disadvantages are the same as for fluorescence lifetime except that there are more examples in the scientific literature of use.

Static Light Scattering

Static light scattering allows monitoring of the sizes of the species in solution. Since proteins typically aggregate upon denaturation (or form fibrils) the detected species size will go up.

This is label-free and independent of specific residues in the protein or buffer composition. The only requirement is that the protein actually aggregates/fibrillates after denaturation and that the protein of interest has been purified.

FastPP

In fast parallel proteolysis the researcher adds a thermostable protease (thermolysin) and takes out samples in parallel upon heating in a thermal gradient cycler. Optionally, for instance for lowly expressed proteins, a Western blot is then run to determine at what temperature a protein becomes degraded. For pure or highly enriched proteins, direct SDS-PAGE detection is possible facilitating Commassie-fluorescence based direct quantification. FastPP exploits that proteins become increasingly susceptible to proteolysis when unfolded and that thermolysin cleaves at hydrophobic residues which are typically found in the core of proteins.

To reduce the workload, Western Blots could be replaced by SDS-PAGE gel polyhistidine-tag staining, provided that the protein has such a tag and is expressed in adequate amounts.

FastPP can be used on unpurified, complex mixtures of proteins and proteins fused with other proteins, such as GST or GFP, as long as the sequence that is the target of the Western blot, e.g. His-tag, is directly linked to the protein of interest. However, commercially available thermolysin is dependent on calcium ions for activity and denatures itself just above 85 degrees Celsius. So calcium must be present and calcium chelators absent in the buffer - other compounds that interfere with the function (such as high concentrations of detegents) of the protease could also be problematic.

FASTpp has also been used to monitor binding-coupled folding of intrinsically disordered proteins (IDPs).

CETSA

CETSA stands for cellular thermal shift assay. Very similar to fastPP but instead of adding a protease samples are centrifuged after heating and the supernatant is run on Western blots or two target-directed antibodies are added and their relative proximity is detected through FRET. When a protein denatures it aggregates and becomes part of the pellet and can therefore not be found in the supernatant. An intrinsic limitation of this assay is that not all proteins aggregate upon unfolding but might instead populate highly soluble, relatively compact molten-globule (like) conformations. In addition, it is possible that proteins will co-precipitate with their less stable protein interaction partners and therefore show lower apparent stability than their actual thermodynamic stability.

Just as for FastPP the Western blot could be replaced by staining for a polyhistidine-tag in the SDS-PAGE gel. Additionally, instead of using two antibodies the FRET method could be performed with a single fluorescent antibody and a fluorescently labelled peptide recognised by the antibody.

A technique that has been shown to work on both complex mixtures and whole cells but unfortunately requires antibodies. The protein of interest should not be fused to other proteins such as MBP or GFP as they could keep a denatured protein in solution.

A CETSA assay for membrane proteins was described recently. This method not only allows to screen ligands but also thermostabilizing conditions.

ThermoFAD

Thermofluor variant specific for flavin-binding proteins. Analogous to thermofluor binding assays, a small volume of protein solution is heated up and the fluorescence increase is followed as function of temperature. In contrast to Thermofluor, no external fluorescent dye is needed because the flavin cofactor is already present in the flavin-binding protein and its fluorescence properties change upon unfolding.

SEC-TS

Size exclusion chromatography can be used directly to access protein stability in the presence or absence of ligands. Samples of purified protein are heated in a water bath or thermocycler, cooled, centrifuged to remove aggregated proteins, and run on an analytical HPLC. As the melting temperature is reached and protein precipitates or aggregates, peak height decreases and void peak height increases. This can be used to identify ligands and inhibitors, and optimize purification conditions.

While of lower through-put than FSEC-TS, requiring large amounts of purified protein, SEC-TS avoids any influence of the fluorescent tag on apparent protein stability.

FSEC-TS

In fluorescence-detection size exclusion chromatography the protein of interest is fluorescently tagged, e.g. with GFP, and run through a gel filtration column on an FPLC system equipped with a fluorescence detector. The resulting chromatogram allows allows the researcher to estimate the monodispersity and expression level of the tagged protein in the current buffer. Since only fluorescence is measured, only the tagged protein is seen in the chromatogram. FSEC is typically used to compare membrane protein orthologs or screen detergents to solubilise specific membrane proteins in.

For fluorescence-detection size-exclusion chromatography-based thermostability assay (FSEC-TS) the samples are heated in the same manner as in FastPP and CETSA and after centrifugation to clear away precipitate the supernatant is treated in the same manner as FSEC. Larger aggregates are seen in the void volume while the peak height for the protein of interest decreases when the unfolding temperature is reached.

GFP has a T_m of ~76 C so the technique is limited to temperature below ~70 C. Of all the techniques mentioned it may have the lowest throughput but it is free of antibody requirements and works on complex mixtures of proteins.

Radioligand Binding Thermostability Assay

GPCRs are pharmacologically important transmembrane proteins. Their x-ray crystal structures were revealed long after other transmembrane proteins of lesser interest. The difficulty in obtaining protein crystals of GPCRs was likely due to their high flexibility. Less flexible versions were obtained by truncating, mutating, and inserting T4 lysozyme in the recombinant sequence. One of the methods researchers used to guide these alterations was radioligand binding thermostability assay.

The assay is performed by incubating the protein with a radiolabelled ligand of the protein for 30 minutes at a given temperature, then quench on ice, run through a gel filtration mini column, and quantify the radiation levels of the protein that comes off the column. The radioligand concentration is high enough to saturate the protein. Denatured protein is unable to bind the radioligand and the protein and radioligand will be separated in the gel filtration mini column. When screening mutants selection will be for thermal stability in the specific conformation, i.e. if the radioligand is an agonist selection will be for the agonist binding conformation and if it is an antagonist then the screening is for stability in the antagonist binding conformation.

Radioassays have the advantage of working with minute amounts of protein. But it is work with radioactive substances and large amount of manual labour is involved. A high-affinity ligand has to be known for the protein of interest and the buffer must not interfere with the binding of the radioligand. Other thermal shift assays can also select for specific conformations if a ligand of the appropriate type is added to the experiment.

ATP Test

The ATP test is a process of rapidly measuring actively growing microorganisms through detection of adenosine triphosphate, or ATP.

ATP Testing Method

ATP is a molecule found in and around living cells, and as such it gives a direct measure of biological concentration and health. ATP is quantified by measuring the light produced through its reaction with the naturally occurring firefly enzyme luciferase using a luminometer. The amount of light produced is directly proportional to the amount of ATP present in the sample.

ATP Tests Can be Used to:

- Control biological treatment reactors.
- Guide biocide dosing programs.
- Determine drinking water cleanliness.
- Manage fermentation processes.
- Assess soil activity.
- Determine corrosion / deposit process type.
- Measure equipment or product sanitation.

1st Generation Testing Vs. 2nd Generation Testing

1st generation ATP tests are derived from hygiene monitoring uses where samples are relatively free of interferences. 2nd Generation tests are specifically designed for water, wastewater and industrial applications where, for the most part, samples contain a variety of components that can interfere with the ATP assay.

How ATP is Measured

ATP is a molecule found only in and around living cells, and as such it gives a direct measure of biological concentration and health. ATP is quantified by measuring the light produced through its reaction with the naturally-occurring firefly enzyme Luciferase using a Luminometer. The amount of light produced is directly proportional to the amount of biological energy present in the sample.

Within a water sample containing microorganisms, there are two types of ATP:

- Intracellular ATP – ATP contained within living biological cells.
- Extracellular ATP – ATP located outside of biological cells that has been released from dead or stressed organisms.

Accurate measurement of these two types of ATP is critical to utilizing ATP-based measurements. Being able to accurately measure these different types of ATP offers the ability to assess biological health and activity, and subsequently control water and wastewater processes.

Phenotypic Testing of Mycobacteria

In microbiology, the phenotypic testing of mycobacteria uses a number of methods. The most-commonly used phenotypic tests to identify and distinguish *Mycobacterium* strains and species from each other are described below.

Tests

Acetamide as sole C and N sources:

Media: KH_2PO_4 (0.5 g), MgSO>$_4$*7H_2o (0.5 g), purified agar (20 g), distilled water (1000 ml). The medium is supplemented with acetamide to a final concentration of 0.02M, adjusted to a pH of 7.0 and sterilized by autoclaving at 115°C for 30 minutes. After sloping, the medium is inoculated with one loop of the cultures and incubated. Growth is read after incubation for two weeks (rapid growers) or four weeks (slow growers).

Arylsulfatase test:

Arylsulfatase enzyme is present in most mycobacteria. The rate by which arylsulfatase enzyme breaks down phenolphthalein disulfate into phenolphthalein (which forms a red color in the presence of sodium bicarbonate) and other salts is used to differentiate certain strains of Mycobacteria. 3 day arylsulfatase test is used to identify potentially pathogenic rapid growers such as M.

fortuitum and M. chelonae. Slow growing M. marinum and M. szulgai are positive in the 14-day arylsulfatase test.

Catalase, semiquantitative activity:

Most mycobacteria produce the enzyme catalase, but they vary in the quantity produced. Also, some forms of catalase are inactivated by heating at 68°C for 20 minutes (others are stable). Organisms producing the enzyme catalase have the ability to decompose hydrogen peroxide into water and free oxygen. The test differs from that used to detect catalase in other types of bacteria by using 30% hydrogen peroxide in a strong detergent solution (10% polysorbate 80).

Citrate:

Sole carbon source.

Egg medium:

Growth on Löwenstein–Jensen medium (LJ medium).

L-Glutamate:

Sole carbon and nitrogen source.

Growth rate:

The growth rate is the length of time required to form mature colonies visible without magnification on solid media. Mycobacteria forming colonies visible to the naked eye within seven days on subculture are known as rapid growers, while those requiring longer periods are termed slow growers.

Iron uptake:

The ability to take up iron from an inorganic iron containing reagent helps differentiate some species of mycobacteria.

Lebek medium:

Lebek is a semisolid medium used to test the oxygen preferences of mycobacterial isolates. Aerophilic growth is indicated by growth on (and above) the surface of the glass wall of the tube; microaerophilic growth is indicated by growth below the surface.

MacConkey agar without crystal violet.

Niacin accumulation (paper strip method):

Niacin is formed as a metabolic byproduct by all mycobacteria, but some species possess an enzyme that converts free niacin to niacin ribonucleotide. *M. tuberculosis* (and some other species) lack this enzyme, and accumulate niacin as a water-soluble byproduct in the culture medium.

Nitrate reduction:

Mycobacteria containing nitroreductase catalyze the reduction from nitrate to nitrite. The presence of nitrite in the test medium is detected by addition of sulfanilamide and n-naphthylethylendiamine. If nitrate is present, red diazonium dye is formed.

Photoreactivity of mycobacteria:

Some mycobacteria produce carotenoid pigments without light; others require photoactivation for pigment production. Photochromogens produce non-pigmented colonies when grown in the dark, and pigmented colonies after exposure to light and re-incubation. Scotochromogens produce deep-yellow-to-orange colonies when grown in either light or darkness. Non-photochromogens are non-pigmented in light and darkness or have a pale-yellow, buff or tan pigment which does not intensify after light exposure.

Picrate tolerance:

Grows on Sauton agar containing picric acid (0.2% w/v) after three weeks

Pigmentation:

Some mycobacteria produce carotenoid pigments without light; others require photoactivation for pigment production.

Pyrazinamide sensitivity (PZA):

The deamidation of pyrazinamide to pyrazinoic acid (assumed to be the active component of the drug pyrazinamide) in four days is a useful physiologic characteristic by which *M. tuberculosis*-complex members can be distinguished.

Sodium chloride tolerance:

Growth on LJ medium containing 5% NaCl

Thiophene-2carboxylic acid hydrazide (TCH) sensitivity:

The growth of *M. bovis* and M. africanum subtype II is inhibited by thiophene-2carboxylic acid hydrazide; growth of *M. tuberculosis* and *M. africanum* subtype I is uninhibited.

Polysorbate 80 hydrolysis:

A test for lipase using polysorbate 80 (polyoxyethylene sorbitan monooleate, a detergent). Certain mycobacteria possess a lipase that splits it into oleic acid and polyoxyethylated sorbitol. The test solution also contains phenol red, which is stabilised by the polysorbate 80; when the latter 80 is hydrolysed, the phenol red changes from yellow to pink.

Urease (adaptation to mycobacteria):

With an inoculation loop, several loopfuls of mycobacteria test colonies are transferred to 0.5 mL of urease substrate, mixed to emulsify and incubated at 35 °C for three days; a colour change (from amber-yellow to pink-red) is sought.

Electroporation

Electroporation, or electropermeabilization, is a microbiology technique in which an electrical field is applied to cells in order to increase the permeability of the cell membrane, allowing chemicals, drugs, or DNA to be introduced into the cell. In microbiology, the process of electroporation is often used to transform bacteria, yeast, or plant protoplasts by introducing new coding DNA. If bacteria and plasmids are mixed together, the plasmids can be transferred into the bacteria after electroporation, though depending on what is being transferred cell-penetrating peptides or CellSqueeze could also be used. Electroporation works by passing thousands of volts across a distance of one to two millimeters of suspended cells in an electroporation cuvette (1.8 - 2.5 kV, 12.5 - 18 kV/cm). Afterwards, the cells have to be handled carefully until they have had a chance to divide, producing new cells that contain reproduced plasmids. This process is approximately ten times more effective than chemical transformation.

Electroporation is also highly efficient for the introduction of foreign genes into tissue culture cells, especially mammalian cells. For example, it is used in the process of producing knockout mice, as well as in tumor treatment, gene therapy, and cell-based therapy. The process of introducing foreign DNA into eukaryotic cells is known as transfection. Electroporation is highly effective for transfecting cells in suspension using electroporation cuvettes. Electroporation has proven efficient for use on tissues in vivo, for in utero applications as well as in ovo transfection. Adherent cells can also be transfected using electroporation, providing researchers with an alternative to trypsinizing their cells prior to transfection.

Laboratory Practice

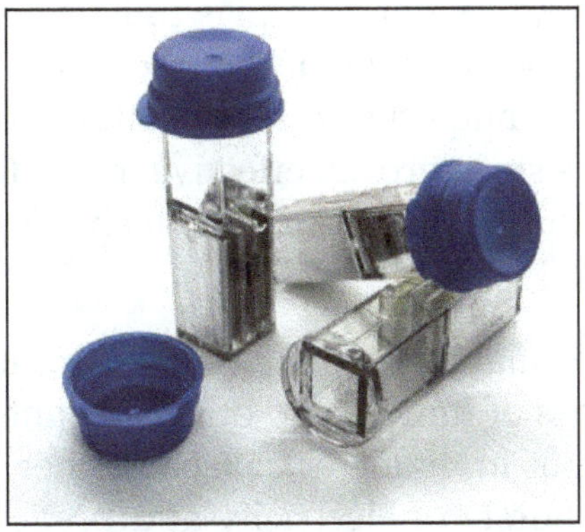

Cuvettes for electroporation. These are plastic with aluminium electrodes and a blue lid. They hold a maximum of 400 μl.

Electroporation is performed with electroporators, purpose-built appliances which create an electrostatic field in a cell solution. The cell suspension is pipetted into a glass or plastic cuvette which has two aluminum electrodes on its sides. For bacterial electroporation, typically a suspension of around 50 microliters is used. Prior to electroporation, this suspension of bacteria is mixed with the plasmid to be transformed. The mixture is pipetted into the cuvette, the voltage and capacitance are set, and the cuvette is inserted into the electroporator. The process requires direct contact between the electrodes and the suspension. Immediately after electroporation, one milliliter of liquid medium is added to the bacteria (in the cuvette or in an Eppendorf tube), and the tube is incubated at the bacteria's optimal temperature for an hour or more to allow recovery of the cells and expression of the plasmid, followed by bacterial culture on agar plates.

The success of the electroporation depends greatly on the purity of the plasmid solution, especially on its salt content. Solutions with high salt concentrations might cause an electrical discharge (known as arcing), which often reduces the viability of the bacteria. For a further detailed investigation of the process, more attention should be paid to the output impedance of the porator device and the input impedance of the cells suspension (e.g. salt content).

Since the cell membrane is not able to pass current (except in ion channels), it acts as an electrical capacitor. Subjecting membranes to a high-voltage electric field results in their temporary breakdown, resulting in pores that are large enough to allow macromolecules (such as DNA) to enter or leave the cell.

Medical Applications

The first research looking at how electroporation might be used on human cells was conducted by researchers at Eastern Virginia Medical School and Old Dominion University, and published in 2003.

The first successful treatment of malignant cutaneous tumors implanted in mice was completed in 2007 by a group of scientists who achieved complete tumor ablation in 12 out of 13 mice. They accomplished this by sending 80 pulses of 100 microseconds at 0.3 Hz with an electrical field magnitude of 2500 V/cm to treat the cutaneous tumors.

A higher voltage of electroporation was found in pigs to irreversibly destroy target cells within a narrow range while leaving neighboring cells unaffected, and thus represents a promising new treatment for cancer, heart disease and other disease states that require removal of tissue. Irreversible electroporation (IRE) has since proven effective in treating human cancer, with surgeons at Johns Hopkins and other institutions now using the technology to treat pancreatic cancer previously thought to be unresectable.

A recent technique called non-thermal irreversible electroporation (N-TIRE) has proven successful in treating many different types of tumors and other unwanted tissue. This procedure is done using small electrodes (about 1mm in diameter), placed either inside or surrounding the target tissue to apply short, repetitive bursts of electricity at a predetermined voltage and frequency. These bursts of electricity increase the resting transmembrane potential (TMP), so that nanopores form in the plasma membrane. When the electricity applied to the tissue is above the electric field threshold of the target tissue, the cells become permanently permeable from the formation of nanopores. As a result, the cells are unable to repair the damage and die due to a loss of homeostasis. N-TIRE is unique to other tumor ablation techniques in that it does not create thermal damage to the tissue around it.

Contrastingly, reversible electroporation occurs when the electricity applied with the electrodes is below the electric field threshold of the target tissue. Because the electricity applied is below the cells' threshold, it allows the cells to repair their phospholipid bilayer and continue on with their normal cell functions. Reversible electroporation is typically done with treatments that involve getting a drug or gene (or other molecule that is not normally permeable to the cell membrane) into the cell. Not all tissue has the same electric field threshold; therefore careful calculations need to be made prior to a treatment to ensure safety and efficacy.

One major advantage of using N-TIRE is that, when done correctly according to careful calculations, it only affects the target tissue. Proteins, the extracellular matrix, and critical structures such as blood vessels and nerves are all unaffected and left healthy by this treatment. This allows for a quicker recovery, and facilitates a more rapid replacement of dead tumor cells with healthy cells.

Before doing the procedure, scientists must carefully calculate exactly what needs to be done, and treat each patient on an individual case-by-case basis. To do this, imaging technology such as CT scans and MRI's are commonly used to create a 3D image of the tumor. From this information, they can approximate the volume of the tumor and decide on the best course of action including the insertion site of electrodes, the angle they are inserted in, the voltage needed, and more, using software technology. Often, a CT machine will be used to help with the placement of electrodes during the procedure, particularly when the electrodes are being used to treat tumors in the brain.

The entire procedure is very quick, typically taking about five minutes. The success rate of these procedures is high and is very promising for future treatment in humans. One disadvantage to using N-TIRE is that the electricity delivered from the electrodes can stimulate muscle cells to contract, which could have lethal consequences depending on the situation. Therefore, a paralytic agent must be used when performing the procedure. The paralytic agents that have been used in such research are successful; however, there is always some risk, albeit slight, when using anesthetics.

A more recent technique has been developed called high-frequency irreversible electroporation (H-FIRE). This technique uses electrodes to apply bipolar bursts of electricity at a high frequency, as opposed to unipolar bursts of electricity at a low frequency. This type of procedure has the same tumor ablation success as N-TIRE. However, it has one distinct advantage, H-FIRE does not cause muscle contraction in the patient and therefore there is no need for a paralytic agent.

Drug and Gene Delivery

Electroporation can also be used to help deliver drugs or genes into the cell by applying short and intense electric pulses that transiently permeabilize cell membrane, thus allowing transport of molecules otherwise not transported through a cellular membrane. This procedure is referred to as electrochemotherapy when the molecules to be transported are chemotherapeutic agents or gene electrotransfer when the molecule to be transported is DNA. Scientists from Karolinska Institutet and the University of Oxford use electroporation of exosomes to deliver siRNAs, antisense oligonucleotides, chemotherapeutic agents and proteins specifically to neurons after inject them systemically (in blood). Because these exosomes are able to cross the blood brain barrier this protocol could solve the issue of poor delivery of medications to the central nervous system and cure Alzheimer's, Parkinson's Disease and brain cancer among other diseases.

Physical Mechanism

Electroporation allows cellular introduction of large highly charged molecules such as DNA which would never passively diffuse across the hydrophobic bilayer core. This phenomenon indicates that the mechanism is the creation of nm-scale water-filled holes in the membrane. Although electroporation and dielectric breakdown both result from application of an electric field, the mechanisms involved are fundamentally different. In dielectric breakdown the barrier material is ionized, creating a conductive pathway. The material alteration is thus chemical in nature. In contrast, during electroporation the lipid molecules are not chemically altered but simply shift position, opening up a pore which acts as the conductive pathway through the bilayer as it is filled with water.

Electroporation is a dynamic phenomenon that depends on the local transmembrane voltage at each point on the cell membrane. It is generally accepted that for a given pulse duration and shape, a specific transmembrane voltage threshold exists for the manifestation of the electroporation phenomenon (from 0.5 V to 1 V). This leads to the definition of an electric field magnitude threshold for electroporation (E_{th}). That is, only the cells within areas where $E \geqq E_{th}$ are electroporated. If a second threshold (E_{ir}) is reached or surpassed, electroporation will compromise the viability of the cells, *i.e.*, irreversible electroporation (IRE).

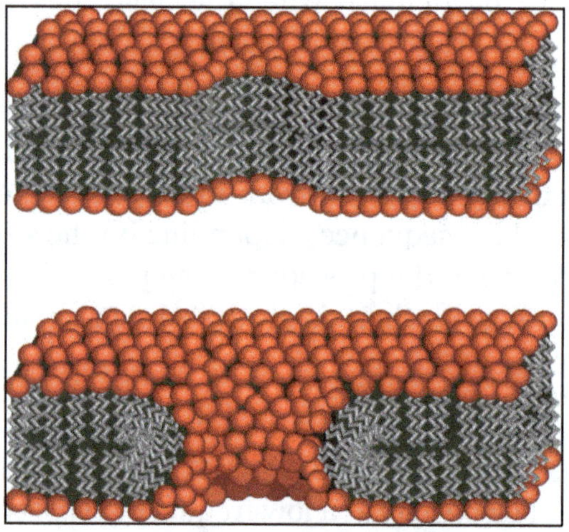

Schematic showing the theoretical arrangement of lipids in a hydrophobic pore (top) and a hydrophilic pore (bottom).

Electroporation is a multi-step process with several distinct phases. First, a short electrical pulse must be applied. Typical parameters would be 300–400 mV for < 1 ms across the membrane (note- the voltages used in cell experiments are typically much larger because they are being applied across large distances to the bulk solution so the resulting field across the actual membrane is only a small fraction of the applied bias). Upon application of this potential the membrane charges like a capacitor through the migration of ions from the surrounding solution. Once the critical field is achieved there is a rapid localized rearrangement in lipid morphology. The resulting structure is believed to be a "pre-pore" since it is not electrically conductive but leads rapidly to the creation of a conductive pore. Evidence for the existence of such pre-pores comes mostly from the "flickering" of pores, which suggests a transition between conductive and insulating states. It has been suggested that these pre-pores are small (~3 Å) hydrophobic defects. If this theory is correct, then the transition to a conductive state could be explained by a rearrangement at the pore edge, in which the lipid heads fold over to create a hydrophilic interface. Finally, these conductive pores can either heal, resealing the bilayer or expand, eventually rupturing it. The resultant fate depends on whether the critical defect size was exceeded which in turn depends on the applied field, local mechanical stress and bilayer edge energy.

Phage Display

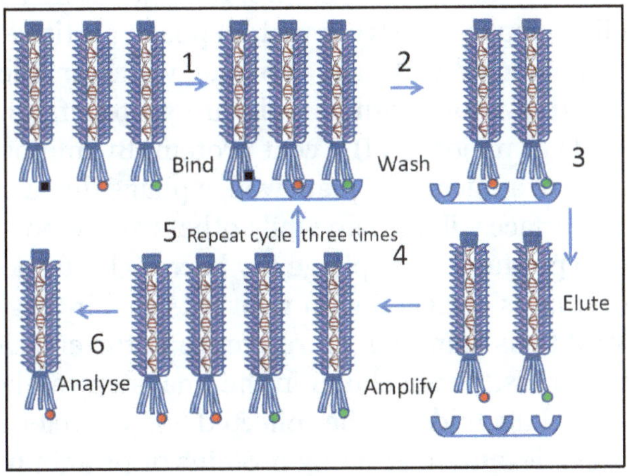

The sequence of events that are followed in phage display screening to identify polypeptides that bind with high affinity to desired target protein or DNA sequence.

Phage display is a laboratory technique for the study of protein–protein, protein–peptide, and protein–DNA interactions that uses bacteriophages (viruses that infect bacteria) to connect proteins with the genetic information that encodes them. In this technique, a gene encoding a protein of interest is inserted into a phage coat protein gene, causing the phage to "display" the protein on its outside while containing the gene for the protein on its inside, resulting in a connection between genotype and phenotype. These displaying phages can then be screened against other proteins, peptides or DNA sequences, in order to detect interaction between the displayed protein and those other molecules. In this way, large libraries of proteins can be screened and amplified in a process called *in vitro* selection, which is analogous to natural selection.

The most common bacteriophages used in phage display are M13 and fd filamentous phage, though T4, T7, and λ phage have also been used.

History

Phage display was first described by George P. Smith in 1985, when he demonstrated the display of peptides on filamentous phage by fusing the peptide of interest onto gene III of filamentous phage. A patent by George Pieczenik claiming priority from 1985 also describes the generation of phage display libraries. This technology was further developed and improved by groups at the Laboratory of Molecular Biology with Greg Winter and John McCafferty, The Scripps Research Institute with Lerner and Barbas and the German Cancer Research Center with Breitling and Dübel for display of proteins such as antibodies for therapeutic protein engineering.

Principle

Like the two-hybrid system, phage display is used for the high-throughput screening of protein interactions. In the case of M13 filamentous phage display, the DNA encoding the protein or peptide of interest is ligated into the pIII or pVIII gene, encoding either the minor or major coat protein, respectively. Multiple cloning sites are sometimes used to ensure that the fragments are inserted in all three possible reading frames so that the cDNA fragment is translated in the proper frame. The phage gene and insert DNA hybrid is then inserted (a process known as "transduction") into *Escherichia coli* (E. coli) bacterial cells such as TG1, SS320, ER2738, or XL1-Blue *E. coli*. If a "phagemid" vector is used (a simplified display construct vector) phage particles will not be released from the *E. coli* cells until they are infected with helper phage, which enables packaging of the phage DNA and assembly of the mature virions with the relevant protein fragment as part of their outer coat on either the minor (pIII) or major (pVIII) coat protein. By immobilizing a relevant DNA or protein target(s) to the surface of a microtiter plate well, a phage that displays a protein that binds to one of those targets on its surface will remain while others are removed by washing. Those that remain can be eluted, used to produce more phage (by bacterial infection with helper phage) and so produce a phage mixture that is enriched with relevant (i.e. binding) phage. The repeated cycling of these steps is referred to as 'panning', in reference to the enrichment of a sample of gold by removing undesirable materials. Phage eluted in the final step can be used to infect a suitable bacterial host, from which the phagemids can be collected and the relevant DNA sequence excised and sequenced to identify the relevant, interacting proteins or protein fragments.

The use of a helper phage can be eliminated by using 'bacterial packaging cell line' technology.

Elution can be done combining low-pH elution buffer with sonification, which, in addition to loosening the peptide-target interaction, also serves to detach the target molecule from the immobilization surface. This ultrasound-based method enables single-step selection of a high-affinity peptide.

Applications

Applications of phage display technology include determination of interaction partners of a protein (which would be used as the immobilised phage "bait" with a DNA library consisting of all coding sequences of a cell, tissue or organism) so that the function or the mechanism of the function of that protein may be determined. Phage display is also a widely used method for *in vitro* protein evolution (also called protein engineering). As such, phage display is a useful tool in drug discovery. It is used for finding new ligands (enzyme inhibitors, receptor agonists and antagonists) to target proteins. The technique is also used to determine tumour antigens (for use in diagnosis and therapeutic targeting) and in searching for protein-DNA interactions using specially-constructed DNA libraries with randomised segments.

Competing methods for *in vitro* protein evolution include yeast display, bacterial display, ribosome display, and mRNA display.

Antibody maturation in vitro

The invention of antibody phage display revolutionised antibody drug discovery. Initial work was done by laboratories at the MRC Laboratory of Molecular Biology (Greg Winter and John McCafferty), the Scripps Research Institute (Richard Lerner and Carlos F. Barbas) and the German Cancer Research Centre (Frank Breitling and Stefan Dübel). In 1991, The Scripps group reported the first display and selection of human antibodies on phage. This initial study described the rapid isolation of human antibody Fab that bound tetanus toxin and the method was then extended to rapidly clone human anti-HIV-1 antibodies for vaccine design and therapy.

Phage display of antibody libraries has become a powerful method for both studying the immune response as well as a method to rapidly select and evolve human antibodies for therapy. Antibody phage display was later used by Carlos F. Barbas at The Scripps Research Institute to create synthetic human antibody libraries, a principle first patented in 1990 by Breitling and coworkers (Patent CA 2035384), thereby allowing human antibodies to be created in vitro from synthetic diversity elements.

Antibody libraries displaying millions of different antibodies on phage are often used in the pharmaceutical industry to isolate highly specific therapeutic antibody leads, for development into antibody drugs primarily as anti-cancer or anti-inflammatory therapeutics. One of the most successful was HUMIRA (adalimumab), discovered by Cambridge Antibody Technology as D2E7 and developed and marketed by Abbott Laboratories. HUMIRA, an antibody to TNF alpha, was the world's first fully human antibody, which achieved annual sales exceeding $1bn.

General Protocol

Below is the sequence of events that are followed in phage display screening to identify polypeptides that bind with high affinity to desired target protein or DNA sequence:

1. Target proteins or DNA sequences are immobilised to the wells of a microtiter plate.
2. Many genetic sequences are expressed in a bacteriophage library in the form of fusions with the bacteriophage coat protein, so that they are displayed on the surface of the viral particle. The protein displayed corresponds to the genetic sequence within the phage.
3. This phage-display library is added to the dish and after allowing the phage time to bind, the dish is washed.
4. Phage-displaying proteins that interact with the target molecules remain attached to the dish, while all others are washed away.
5. Attached phage may be eluted and used to create more phage by infection of suitable bacterial hosts. The new phage constitutes an enriched mixture, containing considerably less irrelevant phage (i.e. non-binding) than were present in the initial mixture.
6. Steps 3 to 6 are optionally repeated one or more times, further enriching the phage library in binding proteins.
7. Following further bacterial-based amplification, the DNA within in the interacting phage is sequenced to identify the interacting proteins or protein fragments.

Selection of the Coat Protein

Filamentous Phages

pIII

pIII is the protein that determines the infectivity of the virion. pIII is composed of three domains (N1, N2 and CT) connected by glycine-rich linkers. The N2 domain binds to the F pilus during virion infection freeing the N1 domain which then interacts with a TolA protein on the surface of the bacterium. Insertions within this protein are usually added in position 249 (within a linker region between CT and N2), position 198 (within the N2 domain) and at the N-terminus (inserted between the N-terminal secretion sequence and the N-terminus of pIII). However, when using the BamHI site located at position 198 one must be careful of the unpaired Cysteine residue (C201) that could cause problems during phage display if one is using a non-truncated version of pIII.

An advantage of using pIII rather than pVIII is that pIII allows for monovalent display when using a phagemid (Ff-phage derived plasmid) combined with a helper phage. Moreover, pIII allows for the insertion of larger protein sequences (>100 amino acids) and is more tolerant to it than pVIII. However, using pIII as the fusion partner can lead to a decrease in phage infectivity leading to problems such as selection bias caused by difference in phage growth rate or even worse, the phage's inability to infect its host. Loss of phage infectivity can be avoided by using a phagemid plasmid and a helper phage so that the resultant phage contains both wild type and fusion pIII.

cDNA has also been analyzed using pIII via a two complementary leucine zippers system, Direct Interaction Rescue or by adding an 8-10 amino acid linker between the cDNA and pIII at the C-terminus.

pVIII

pVIII is the main coat protein of Ff phages. Peptides are usually fused to the N-terminus of pVIII. Usually peptides that can be fused to pVIII are 6-8 amino acids long. The size restriction seems to have less to do with structural impediment caused by the added section and more to do with the size exclusion caused by pIV during coat protein export. Since there are around 2700 copies of the protein on a typical phages, it is more likely that the protein of interest will be expressed polyvalently even if a phagemid is used. This makes the use of this protein unfavorable for the discovery of high affinity binding partners.

To overcome the size problem of pVIII, artificial coat proteins have been designed. An example is Weiss and Sidhu's inverted artificial coat protein (ACP) which allows the display of large proteins at the C-terminus. The ACP's could display a protein of 20kDa, however, only at low levels (mostly only monovalently).

pVI

pVI has been widely used for the display of cDNA libraries. The display of cDNA libraries via phage display is an attractive alternative to the yeast-2-hybrid method for the discovery of interacting proteins and peptides due to its high throughput capability. pVI has been used preferentially to pVIII and pIII for the expression of cDNA libraries because one can add the protein of interest to the C-terminus of pVI without greatly affecting pVI's role in phage assembly. This means that the stop codon in the cDNA is no longer an issue. However, phage display of cDNA is always limited by the inability of most prokaryotes in producing post-translational modifications present in eukaryotic cells or by the misfolding of multi-domain proteins.

While pVI has been useful for the analysis of cDNA libraries, pIII and pVIII remain the most utilized coat proteins for phage display.

pVII and pIX

In an experiment in 1995, display of Glutathione S-transferase was attempted on both pVII and pIX and failed. However, phage display of this protein was completed successfully after the addition of a periplasmic signal sequence (pelB or ompA) on the N-terminus. In a recent study, it has been shown that AviTag, FLAG and His could be displayed on pVII without the need of a signal sequence. Then the expression of single chain Fv's (scFv), and single chain T cell receptors (scTCR) were expressed both with and without the signal sequence.

PelB (an amino acid signal sequence that targets the protein to the periplasm where a signal peptidase then cleaves off PelB) improved the phage display level when compared to pVII and pIX fusions without the signal sequence. However, this led to the incorporation of more helper phage genomes rather than phagemid genomes. In all cases, phage display levels were lower than using pIII fusion. However, lower display might be more favorable for the selection of binders due to lower display being closer to true monovalent display. In five out of six occasions, pVII and pIX fusions without pelB was more efficient than pIII fusions in affinity selection assays. The paper even goes on to state that pVII and pIX display platforms may outperform pIII in the long run.

The use of pVII and pIX instead of pIII might also be an advantage because virion rescue may be undertaken without breaking the virion-antigen bond if the pIII used is wild type. Instead, one could cleave in a section between the bead and the antigen to elute. Since the pIII is intact it does not matter whether the antigen remains bound to the phage.

T7 phages

The issue of using Ff phages for phage display is that they require the protein of interest to be translocated across the bacterial inner membrane before they are assembled into the phage. Some proteins cannot undergo this process and therefore cannot be displayed on the surface of Ff phages. In these cases, T7 phage display is used instead. In T7 phage display, the protein to be displayed is attached to the C-terminus of the gene 10 capsid protein of T7.

The disadvantage of using T7 is that the size of the protein that can be expressed on the surface is limited to shorter peptides because large changes to the T7 genome cannot be accommodated like it is in M13 where the phage just makes its coat longer to fit the larger genome within it. However, it can be useful for the production of a large protein library for scFV selection where the scFV is expressed on an M13 phage and the antigens are expressed on the surface of the T7 phage.

Bioinformatics Resources and Tools

Databases and computational tools for mimotopes have been an important part of phage display study. Databases, programs and web servers have been widely used to exclude target-unrelated peptides, characterize small molecules-protein interactions and map protein-protein interactions. Users can use three dimensional structure of a protein and the peptides selected from phage display experiment to map conformational eptiopes. Some of the fast and efficient computational methods are available online (e.g. EpiSearch http://curie.utmb.edu/episearch.html).

References

- Czyzewski, BK; Wang, DN (11 March 2012). "Identification and characterization of a bacterial hydrosulphide ion channel.". Nature. 483 (7390): 494–7. doi:10.1038/nature10881. PMID 22407320
- Lowman HB, Clackson T (2004). "1.3". Phage display: a practical approach. Oxford [Oxfordshire]: Oxford University Press. pp. 10–11. ISBN 0-19-963873-X
- Ashok, Y; Nanekar, R; Jaakola, VP (December 2015). "Defining thermostability of membrane proteins by western blotting.". Protein engineering, design & selection : PEDS. 28 (12): 539–42. doi:10.1093/protein/gzv049. PMID 26384510
- Kreisig, T; Prasse, AA; Zscharnack, K; Volke, D; Zuchner, T (8 July 2014). "His-tag protein monitoring by a fast mix-and-measure immunoassay.". Scientific Reports. 4: 5613. doi:10.1038/srep05613. PMID 25000910
- Madigan, Michael T. (2012). Brock biology of microorganisms (13th ed.). San Francisco: Benjamin Cummings. ISBN 9780321649638

Permissions

All chapters in this book are published with permission under the Creative Commons Attribution Share Alike License or equivalent. Every chapter published in this book has been scrutinized by our experts. Their significance has been extensively debated. The topics covered herein carry significant information for a comprehensive understanding. They may even be implemented as practical applications or may be referred to as a beginning point for further studies.

We would like to thank the editorial team for lending their expertise to make the book truly unique. They have played a crucial role in the development of this book. Without their invaluable contributions this book wouldn't have been possible. They have made vital efforts to compile up to date information on the varied aspects of this subject to make this book a valuable addition to the collection of many professionals and students.

This book was conceptualized with the vision of imparting up-to-date and integrated information in this field. To ensure the same, a matchless editorial board was set up. Every individual on the board went through rigorous rounds of assessment to prove their worth. After which they invested a large part of their time researching and compiling the most relevant data for our readers.

The editorial board has been involved in producing this book since its inception. They have spent rigorous hours researching and exploring the diverse topics which have resulted in the successful publishing of this book. They have passed on their knowledge of decades through this book. To expedite this challenging task, the publisher supported the team at every step. A small team of assistant editors was also appointed to further simplify the editing procedure and attain best results for the readers.

Apart from the editorial board, the designing team has also invested a significant amount of their time in understanding the subject and creating the most relevant covers. They scrutinized every image to scout for the most suitable representation of the subject and create an appropriate cover for the book.

The publishing team has been an ardent support to the editorial, designing and production team. Their endless efforts to recruit the best for this project, has resulted in the accomplishment of this book. They are a veteran in the field of academics and their pool of knowledge is as vast as their experience in printing. Their expertise and guidance has proved useful at every step. Their uncompromising quality standards have made this book an exceptional effort. Their encouragement from time to time has been an inspiration for everyone.

The publisher and the editorial board hope that this book will prove to be a valuable piece of knowledge for students, practitioners and scholars across the globe.

Index

M

N

P

R

S

Z

www.ingramcontent.com/pod-product-compliance
Lightning Source LLC
LaVergne TN
LVHW080718210926
848485LV00003B/487